Cabin Crew Conference 2016

Held at the 19th World Aviation Training Conference & Tradeshow (WATS 2016)

Orlando, Florida, USA
18 - 21 April 2016

ISBN: 978-1-5108-2541-3

Printed from e-media with permission by:

Curran Associates, Inc.
57 Morehouse Lane
Red Hook, NY 12571

Some format issues inherent in the e-media version may also appear in this print version.

Printed by Curran Associates, Inc. (2016)

For permission requests, please contact Halldale Media Inc.
at the address below.

Halldale Media Inc.
735 Primera Blvd.
Suite 220
Lake Mary, Florida 32746
USA

Phone: (407) 322-5605
Fax: (407) 322-5604

www.halldale.com

Additional copies of this publication are available from:

Curran Associates, Inc.
57 Morehouse Lane
Red Hook, NY 12571 USA
Phone: 845-758-0400
Fax: 845-758-2633
Email: curran@proceedings.com
Web: www.proceedings.com

TABLE OF CONTENTS

BREAKOUT SESSION II

DAY 3 – THURSDAY 21ST APRIL 2016

SESSION 7 – SIMULATION TECHNOLOGY UPDATE

SESSION 8 – MOBILE, E-LEARNING AND NEW MEDIA INSIGHTS

NTSB National Transportation Safety Board

General Aviation and the NTSB Most Wanted List

19th Annual World Aviation Training Conference and Tradeshow
Orlando, April 19, 2016

Earl F. Weener, Ph.D.
Member, NTSB

NTSB Mission

The NTSB is an independent US federal agency charged with determining the probable cause(s) of transportation accidents, making recommendations to prevent their recurrence, conducting special studies and investigations, and coordinating resources to assist victims and their families after an accident.

2 NTSB

Topics

- General Aviation Accident Trends
- Most Wanted List – 2016
  - Distraction
  - Substance Impairment
  - Medical Fitness for Duty
  - Fatigue
  - Occupant Protection
  - Recorders
  - Loss of Control

3 NTSB

All GA Accidents

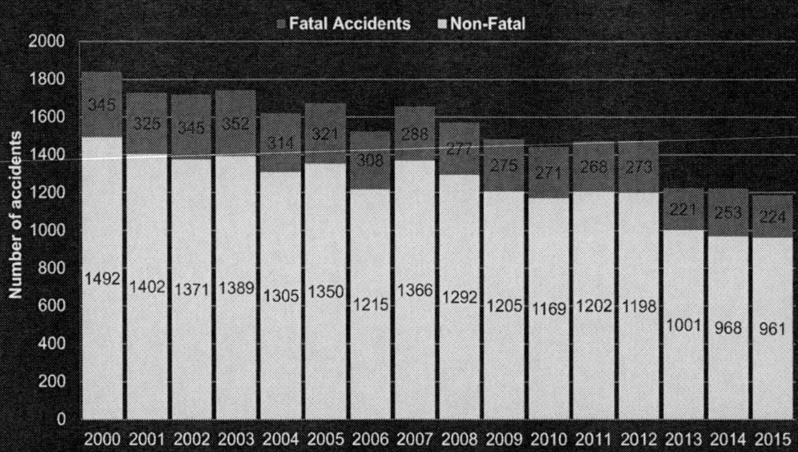

*2015 Preliminary numbers

4 NTSB

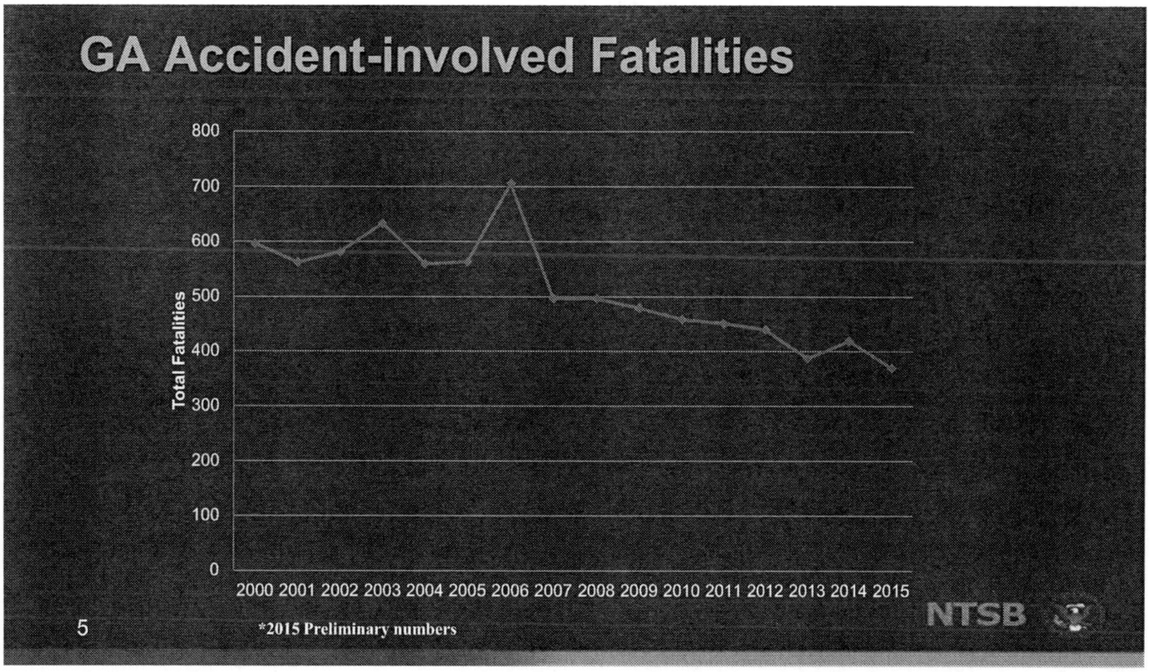
GA Accident-involved Fatalities
800
700
600
500
400
300
200
100
0
Total Fatalities
2000 2001 2002 2003 2004 2005 2006 2007 2008 2009 2010 2011 2012 2013 2014 2015
NTSB
5
*2015 Preliminary numbers

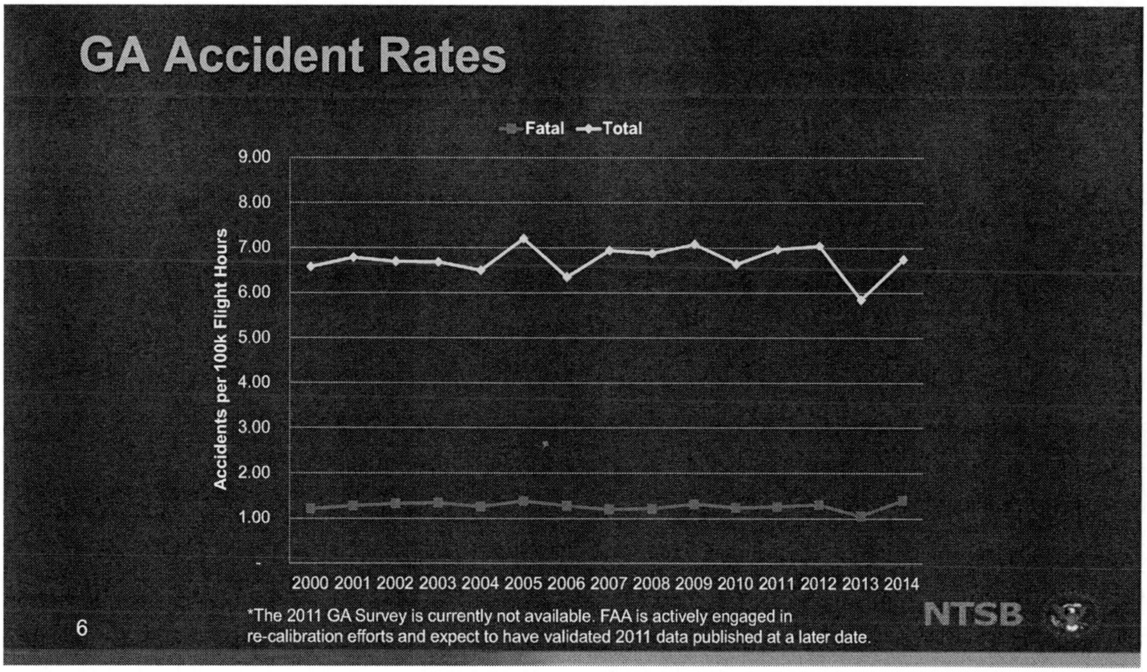
GA Accident Rates
Fatal
Total
9.00
8.00
7.00
6.00
5.00
4.00
3.00
2.00
1.00
-
Accidents per 100k Flight Hours
2000 2001 2002 2003 2004 2005 2006 2007 2008 2009 2010 2011 2012 2013 2014
*The 2011 GA Survey is currently not available. FAA is actively engaged in
re-calibration efforts and expect to have validated 2011 data published at a later date.
NTSB
6

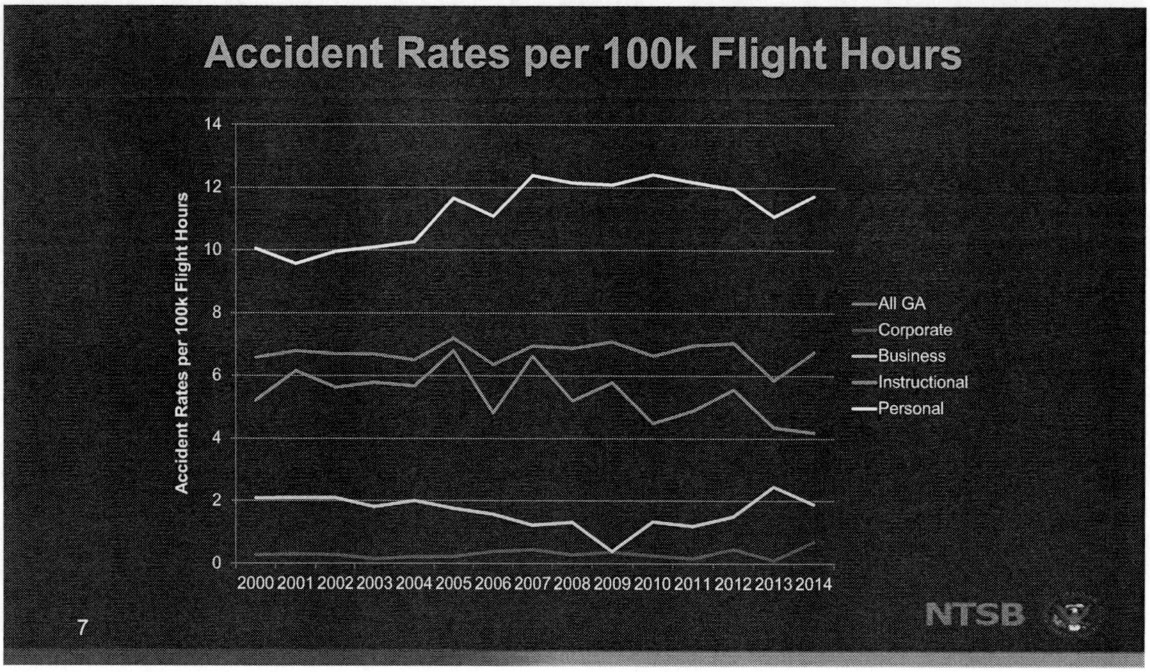
Accident Rates per 100k Flight Hours
Accident Rates per 100k Flight Hours
14
12
10
8
6
4
2
0
2000 2001 2002 2003 2004 2005 2006 2007 2008 2009 2010 2011 2012 2013 2014
All GA
Corporate
Business
Instructional
Personal
7
NTSB

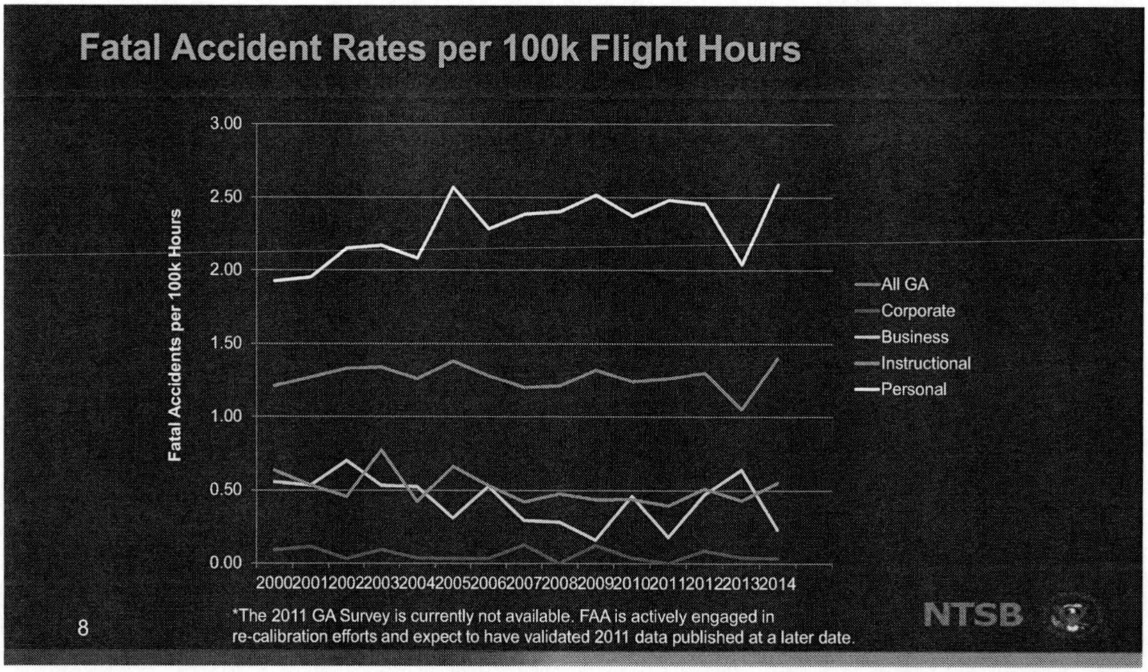
Fatal Accident Rates per 100k Flight Hours
Fatal Accidents per 100k Hours
3.00
2.50
2.00
1.50
1.00
0.50
0.00
200020012002200320042005200620072008200920102011201220132014
All GA
Corporate
Business
Instructional
Personal
*The 2011 GA Survey is currently not available. FAA is actively engaged in re-calibration efforts and expect to have validated 2011 data published at a later date.
8
NTSB

Defining Events – Part 91 Ops

Business	Instruction	Personal
1. Loss of Control	1. Loss of Control	1. Loss of Control
2. CFIT	2. Midair	2. Powerplant Fail
3. Fuel	3. Powerplant Fail	3. CFIT
4. Component Fail	4. CFIT	4. Other
5. Powerplant Fail	5. Other	5. Component Fail

9 NTSB

Why GA on the Most Wanted List?

- NTSB investigates approximately 1500 GA accidents per year over the last decade
- Overall GA accident rate flat
  - Has improved little over the last decade
  - Airline accident rate decreased more than 80%
- Personal flying accident rate
  - Increased 13% over last 10 years
  - Greater than 15% increase in fatals over that period

- ***GA safety needs attention***

10 NTSB

NTSB 2016 Most Wanted List

- Disconnect from Deadly Distractions
- End Substance Impairment in Transportation
- Expand Use of Recorders to Enhance Transportation Safety
- Improve Rail Transit Safety Oversight
- Prevent Loss of Control in Flight in General Aviation
- Promote Availability of Collision Avoidance Technologies in Highway Vehicles
- Promote the Completion of Rail Safety Initiatives
- Reduce Fatigue-Related Accidents
- Require Medical Fitness for Duty
- Strengthen Occupant Protection

11 NTSB

Gray Summit, MO – bus/vehicle/truck crash

12 NTSB

Distraction

13

NTSB

Distraction

14

NTSB

Distraction

15 NTSB

2016 MWL - Distractions

- ***Disconnect from Deadly Distractions***
  - A factor in all modes of transportation
- Aviation emphasis
  - Sterile Cockpit
  - Appropriate use of PEDs
  - Manage distractions

16 NTSB

Impairment

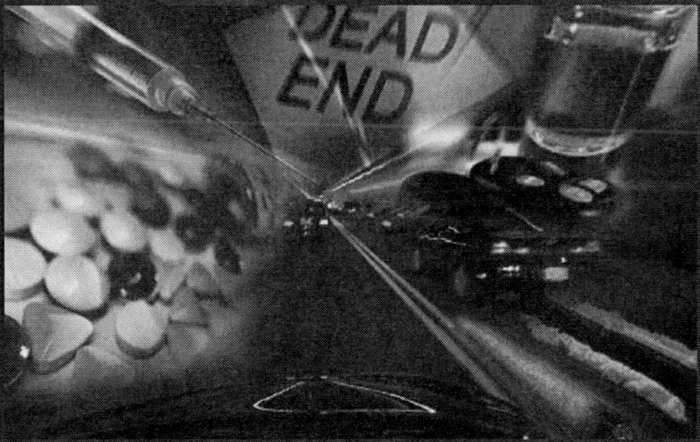

17 NTSB

2016 MWL - Impairment

- ***End Substance Impairment in Transportation***
  - A factor in all modes
- Fatally injured pilots - potentially impairing drugs
  - 11% average 1990 - 1997
  - 23% average 2008 - 2012

18 NTSB

Toxicology Findings by Category, 1990-2012

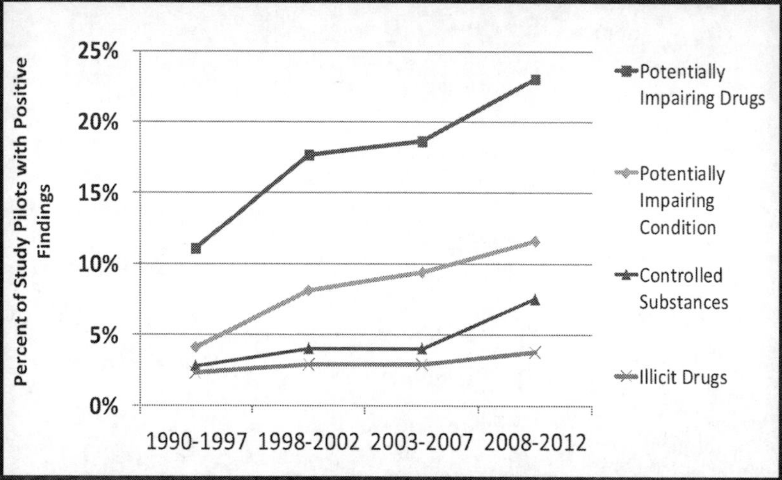

19

Most Common Drugs

- Sedating antihistamines
  - Most common category
- Diphenhydramine
  - Most common individual drug
  - Most common potentially impairing drug
  - Use INCREASING

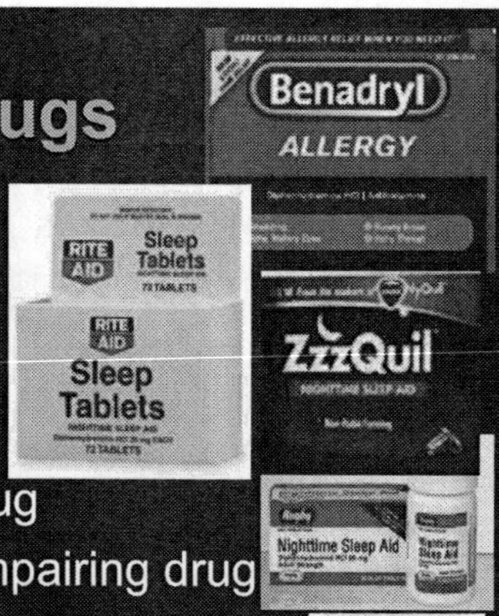

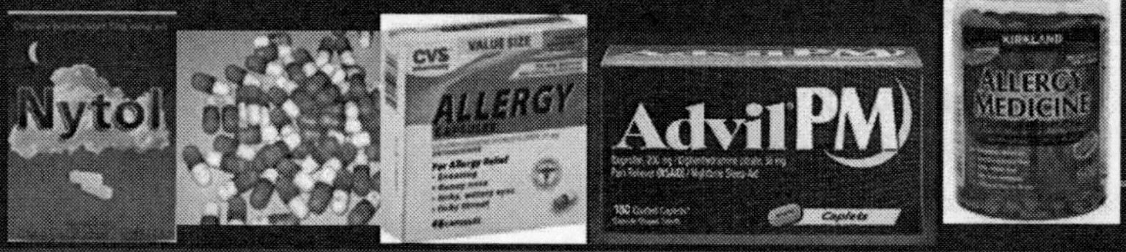

20

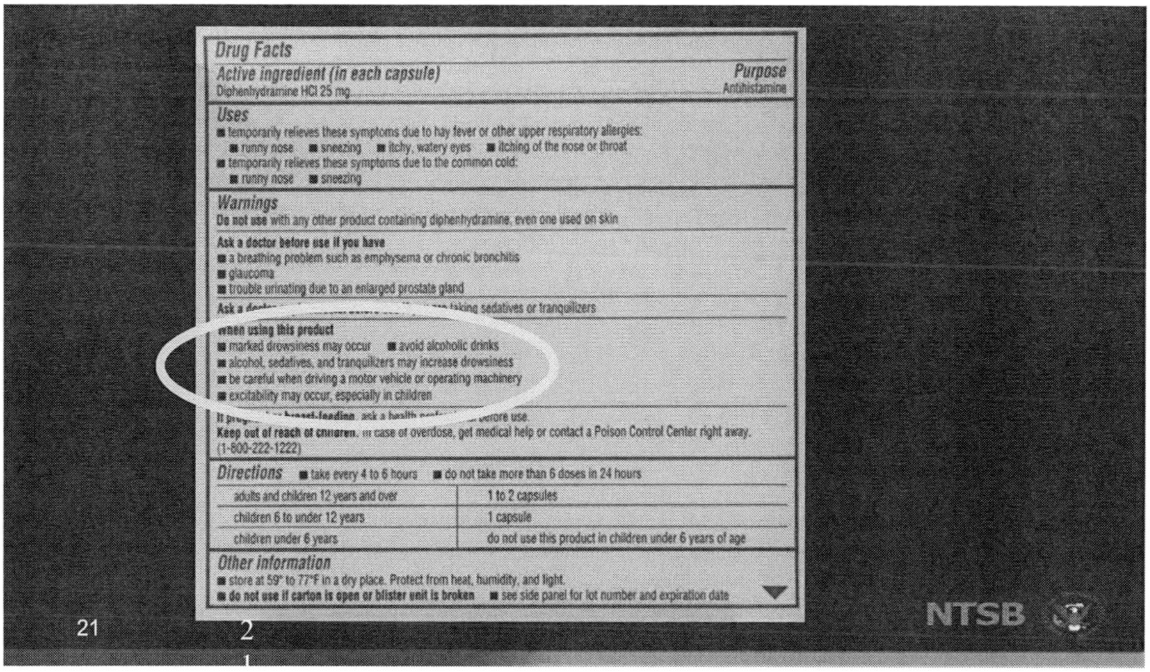

2
1

Medical Resources for Pilots

- General Aviation Joint Steering Committee (GAJSC)
  - 2013 Letter to pilots
  - 2014 Initiatives
    - Drug database
    - Training course

NTSB

22

Medical Resources for Pilots

- FAA Publications
  - *Medications and Flying*
  - *Guide for Aviation Medical Examiners*
- Aircraft Owners and Pilots Association (AOPA)
  - Member resources

23 NTSB

Medical Fitness for Duty

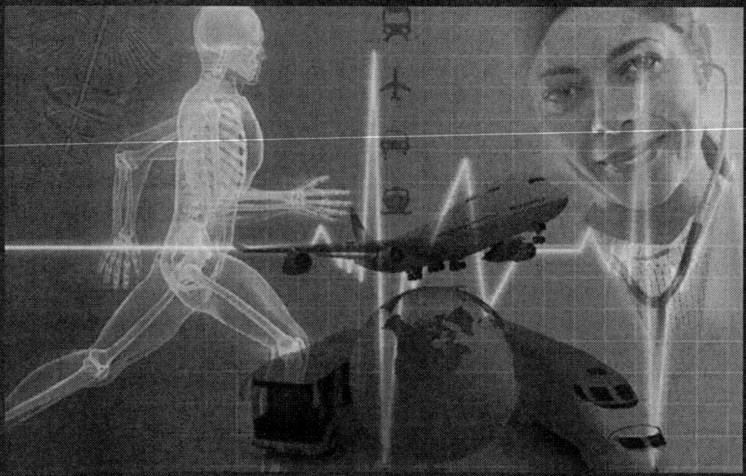

24 NTSB

Fitness for Duty

25

NTSB

2016 MWL – Fitness for Duty

- ***Require Medical Fitness for Duty***
  - A factor in all modes
- Airman Medical – fitness at exam time
- Pilots must self-assess fitness
  - Need for appropriate flight preparations
- Focus on medical conditions
  - Example – Obstructive Sleep Apnea
- Enhance medical knowledge

26

NTSB

Fatigue

27

NTSB

2016 MWL - Fatigue

- ***Reduce Fatigue - Related Accidents***
  - A factor in all modes
  - 182 Major investigations (2001 – 2012)
    - 20% involved fatigue
  - Need
    - Research, education, training
    - Technologies
    - Hours of service, on/off duty policies
    - Medical treatment of sleep disorders

28

NTSB

BE A36 – 2 fatalities/2 survivors

29

NTSB

2016 MWL – Occupant Protection

- ***Strengthen Occupant Protection***
  - A factor in all modes
  - Numerous investigations showed potential for reduced injuries & fatalities
  - Need
    - Increased use of existing restraint systems
    - Designs that enhance survival space
    - Enhance ease of evacuation
    - Shoulder harnesses for GA aircraft

30

NTSB

Recorders

31

NTSB

Recorders

- Expand use of recorders to enhance transportation safety
  - A factor in all modes
  - Critical in accident investigation
  - Install crash resistant image recorders in smaller turbine powered aircraft
  - Install flight recorders in transport category and HEMS aircraft

32

NTSB

Loss of Control

33

NTSB

Loss of Control

34

NTSB

2016 MWL – Loss of Control

- ***Prevent Loss of Control in Flight in General Aviation***
- More than 40% fatal GA accidents were LOC during 2004 – 2014
- Most deadly flight phases
  - Approach to landing
  - Maneuvering
  - Climb

35 NTSB

Business Flying, 2008-2014

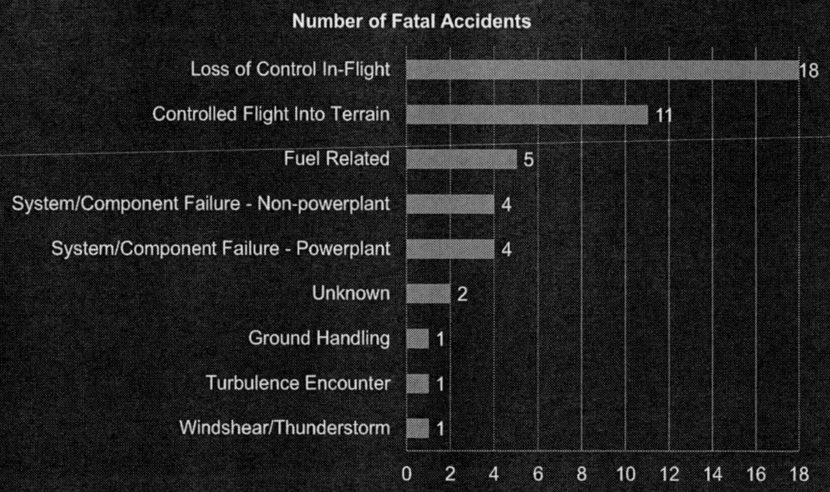

36 NTSB

Instructional Flying, 2008-2014
Number of Fatal Accidents
Loss of Control In-Flight
Midair
System/Component Failure-Powerplant
Controlled Flight Into Terrain
Other
Abrupt Maneuver
Low Altitude Operation
Collision on Takeoff or Landing
Unknown
Abnormal Runway Contact
Ground Handling
Loss of control on Ground
Fuel Related
Simulated/training event
Unintended Flight Into IMC
10
8
7
5
4
4
3
3
2
2
2
1
1
1
0 10 20 30 40 50 60
NTSB
37

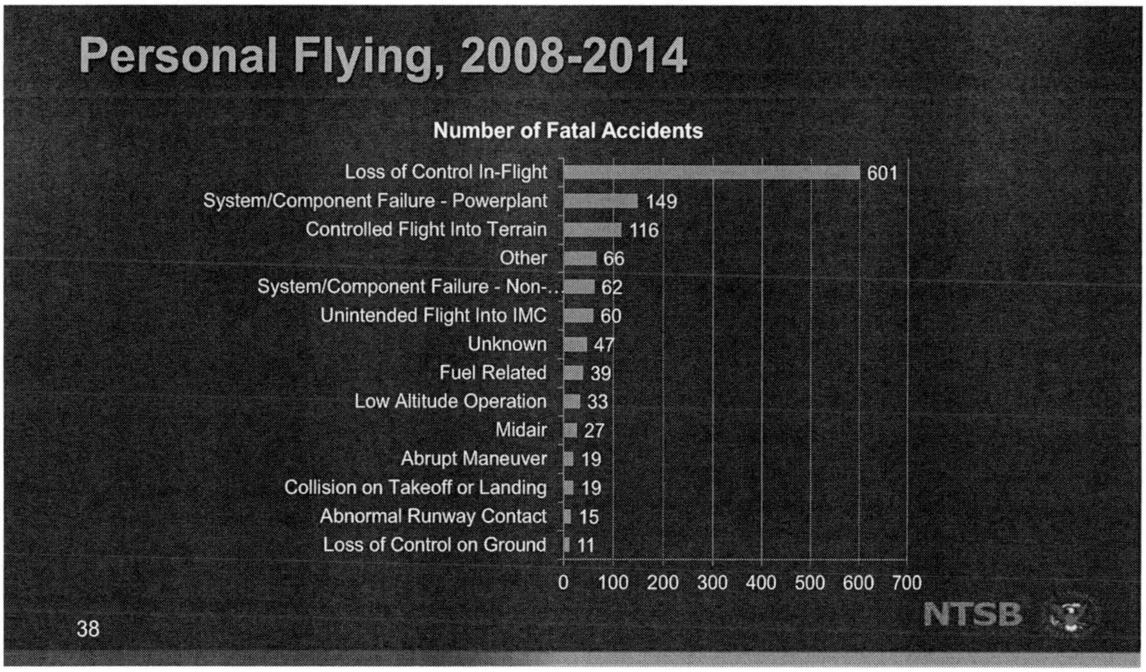
Personal Flying, 2008-2014
Number of Fatal Accidents
Loss of Control In-Flight 601
System/Component Failure - Powerplant 149
Controlled Flight Into Terrain 116
Other 66
System/Component Failure - Non-... 62
Unintended Flight Into IMC 60
Unknown 47
Fuel Related 39
Low Altitude Operation 33
Midair 27
Abrupt Maneuver 19
Collision on Takeoff or Landing 19
Abnormal Runway Contact 15
Loss of Control on Ground 11
0 100 200 300 400 500 600 700
NTSB
38

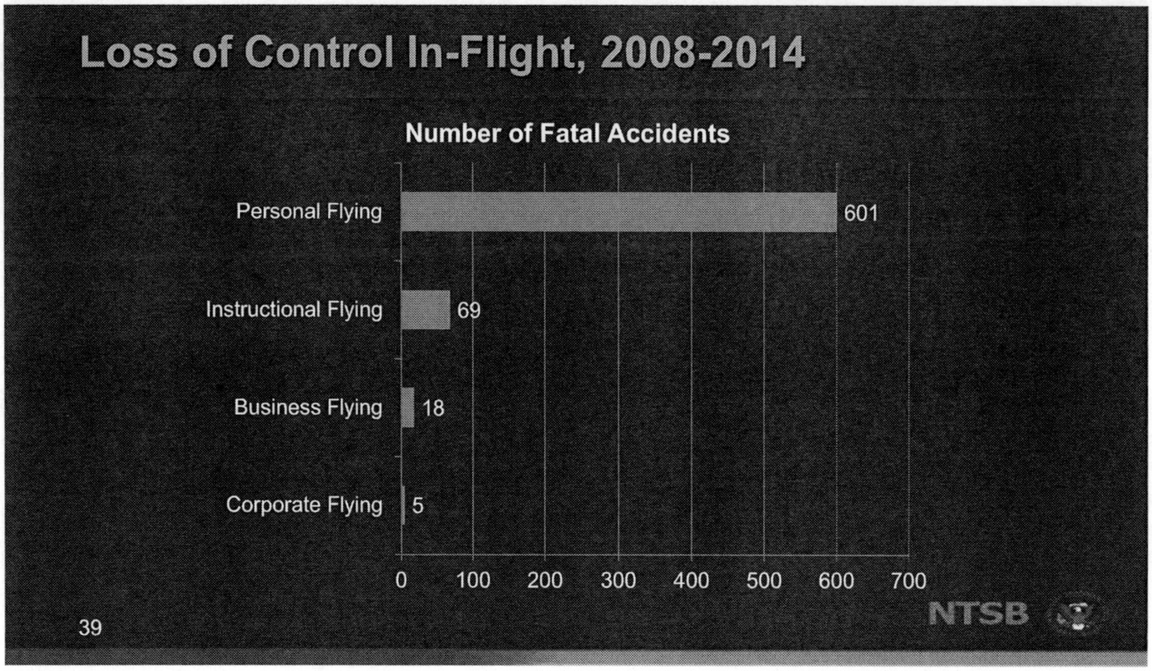
Loss of Control In-Flight, 2008-2014
Number of Fatal Accidents
Personal Flying
601
Instructional Flying
69
Business Flying
18
Corporate Flying
5
0
100
200
300
400
500
600
700
NTSB
39

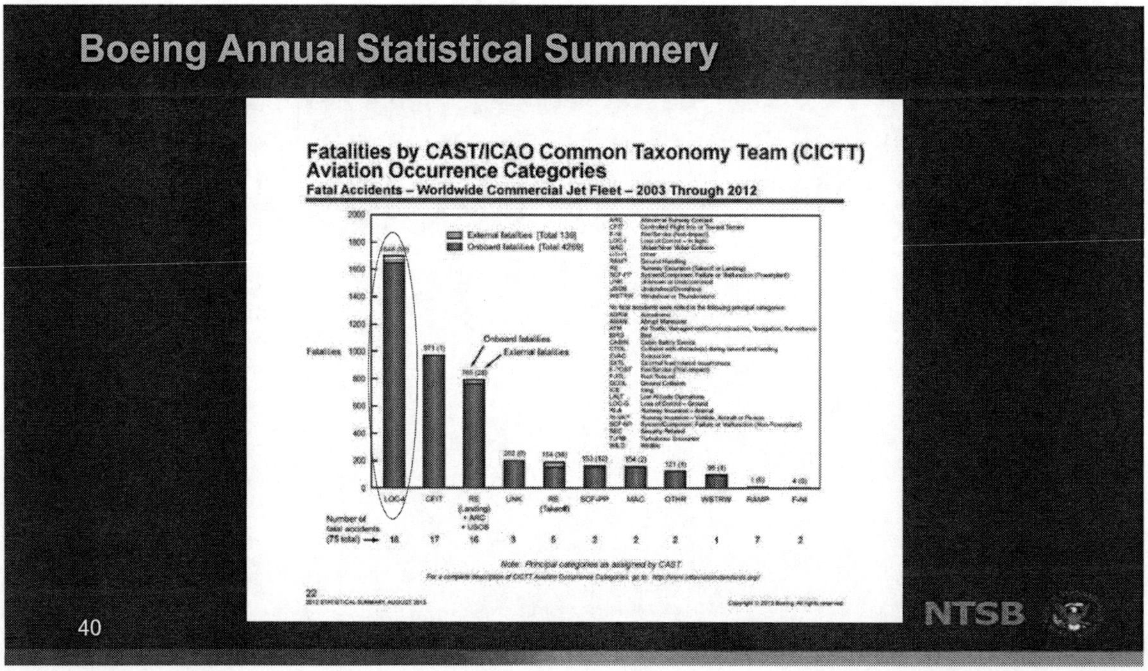
Boeing Annual Statistical Summery
Fatalities by CAST/ICAO Common Taxonomy Team (CICTT) Aviation Occurrence Categories
Fatal Accidents – Worldwide Commercial Jet Fleet – 2003 Through 2012
NTSB
40

Primary category of accidents

Personal flying	– LOC
Instructional flying	– LOC
Business flying	– LOC
Corporate flying	– LOC
Airline flying	– LOC

41 NTSB

Loss-of-control Working Group

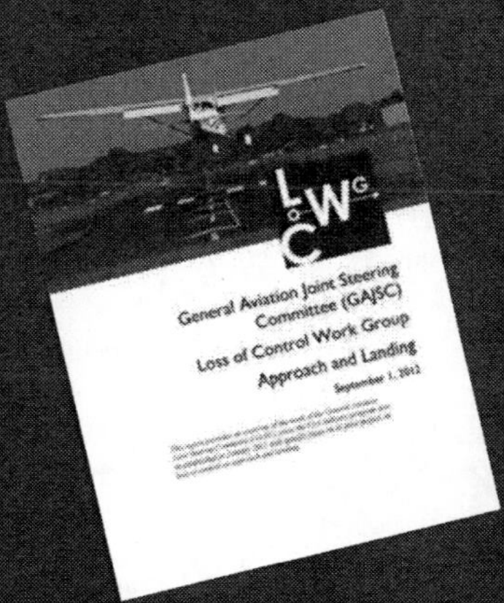

Safety Enhancements Identified

- AOA – New, Current, Retrofit
- Aeronautical Decision Making
- Stabilized Approach
- Single Pilot CRM
- Medication effects
- Weather Technologies
- Etc...

28 Safety Enhancements

42 NTSB

Lower Cost AOA Displays

- Stall occurs at a specific Angle-of-Attack
  - But not necessarily at the same airspeed

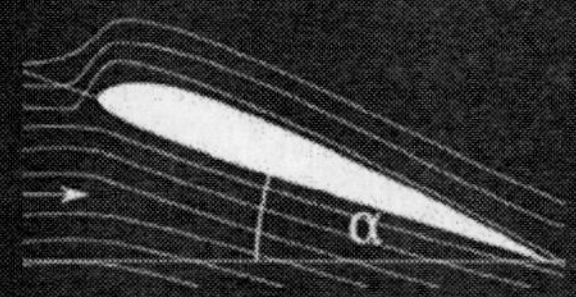

First of AOA indicators built to ASTM stds and installed as a minor mod

FAA installation policy changed

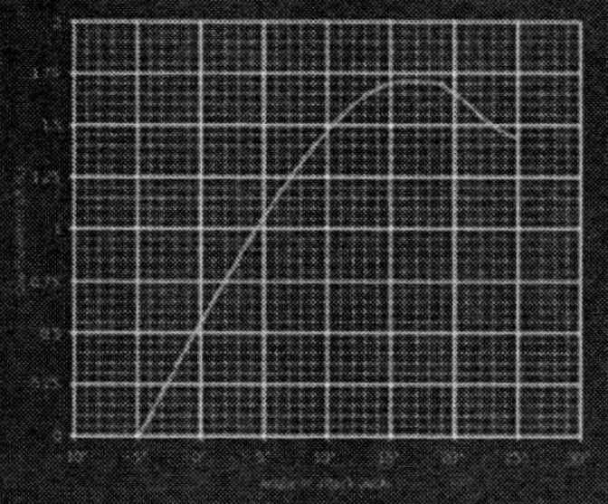

43

NTSB

Stall Recovery

- Reduce the angle-of-attack below stall AOA (Max CL)
  - Push over to eliminate stall warning
- Level wings
- Adjust thrust
  - Avoid overspeed and high G levels
- Pitch back to level flight
- Don’t try to “***Power out of a stall***”

44

NTSB

NTSB LOC Forum – Oct 14, 2015

Humans and Hardware:
Preventing General Aviation
Inflight Loss of Control

- Search "NTSB Loss of Control Forum" and look for Capitol Connection web site.
- Also on YouTube.
- Forum Docket contains 38 documents

45 NTSB

Alfred Sheinwold

"Learn all you can from the mistakes of others. You won't have time to make them all yourself"

46 NTSB

Accident Investigations

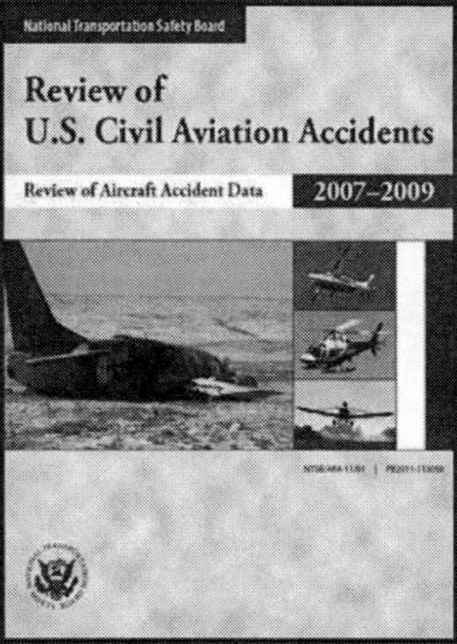

- NTSB accident files are on-line
- Many recent accident Dockets are on-line
  - Factual reports,
  - Interviews
  - Photographs
- www.ntsb.gov

http://www.ntsb.gov/doclib/reports/2011/ARA1101.pdf

47 NTSB

Douglas Adams

"Human beings, who are almost unique in having ability to learn from the experience of others, are also remarkable for their apparent disinclination to do so."

48 NTSB

NATIONAL TRANSPORTATION
SAFETY BOARD
NTSB
49
B

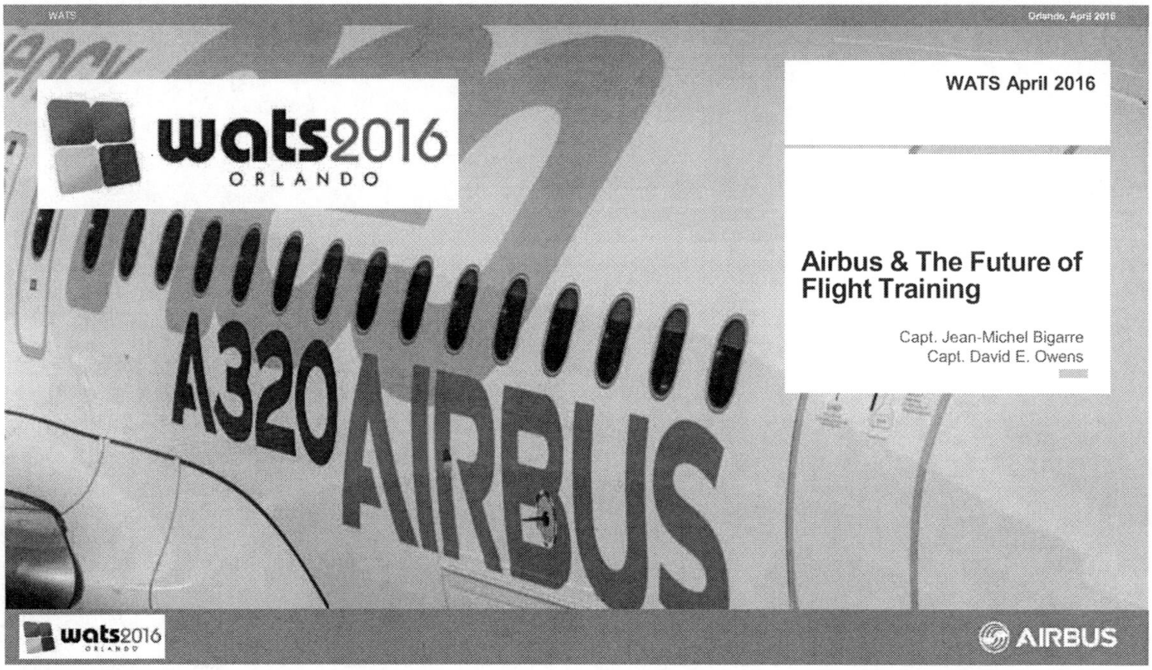

How do we make decisions, how do we train that?

- Going to start with some detail, then build the 'big picture'
- At Airbus we have been taking a fundamental look at training
- Here is one example – Decision Making – why?
- Common reference point in many accidents = 'poor Decision Making'
- Common reference in many 'good' events = 'good Decision Making'
- Here are some of the interesting discoveries we made!
- The first thing we decided was to make it for pilots, from pilots and NO mnemonics!

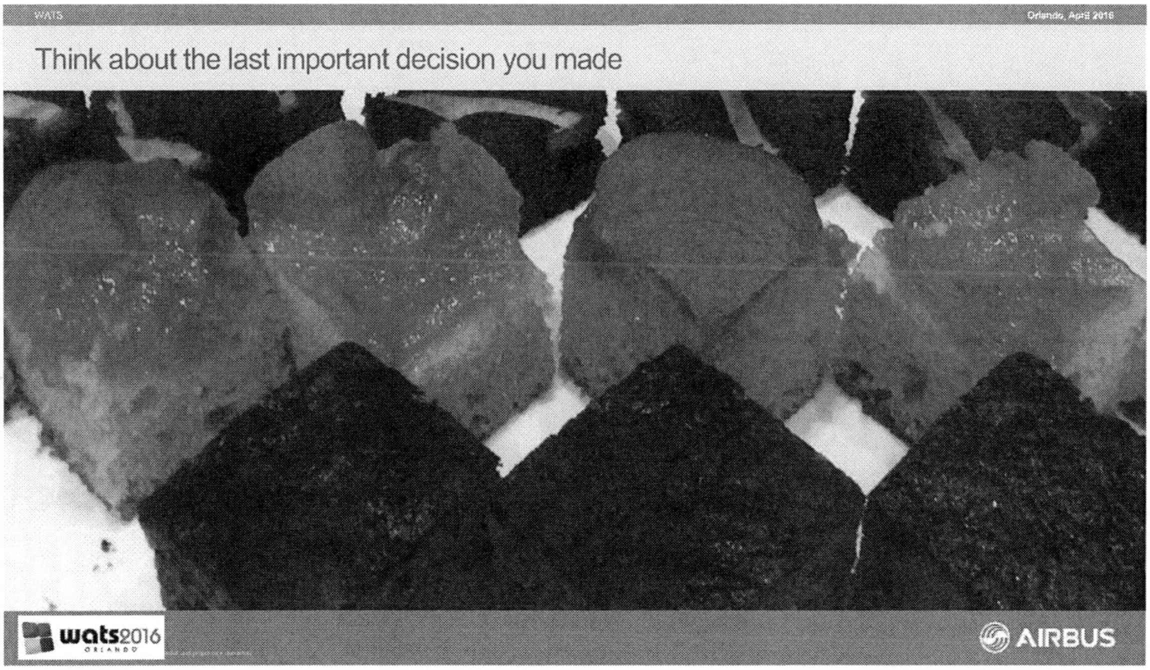
WATS
Orlando, April 2016
Think about the last important decision you made
wats2016
ORLANDO
AIRBUS

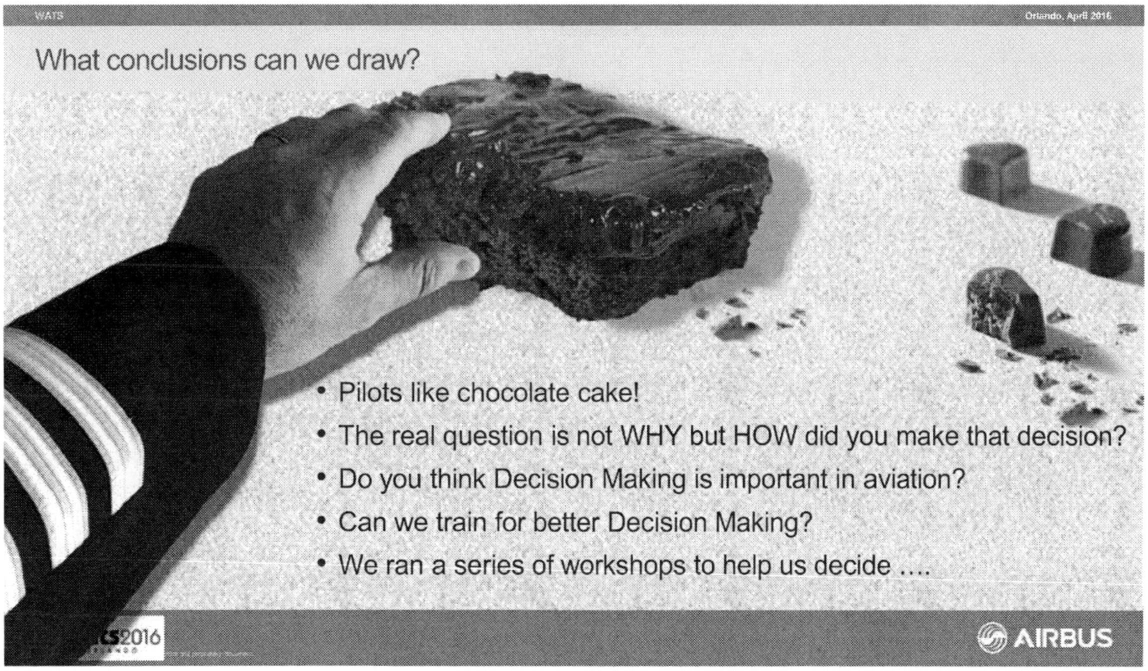
WATS
Orlando, April 2016
What conclusions can we draw?
• Pilots like chocolate cake!
• The real question is not WHY but HOW did you make that decision?
• Do you think Decision Making is important in aviation?
• Can we train for better Decision Making?
• We ran a series of workshops to help us decide ….
AIRBUS

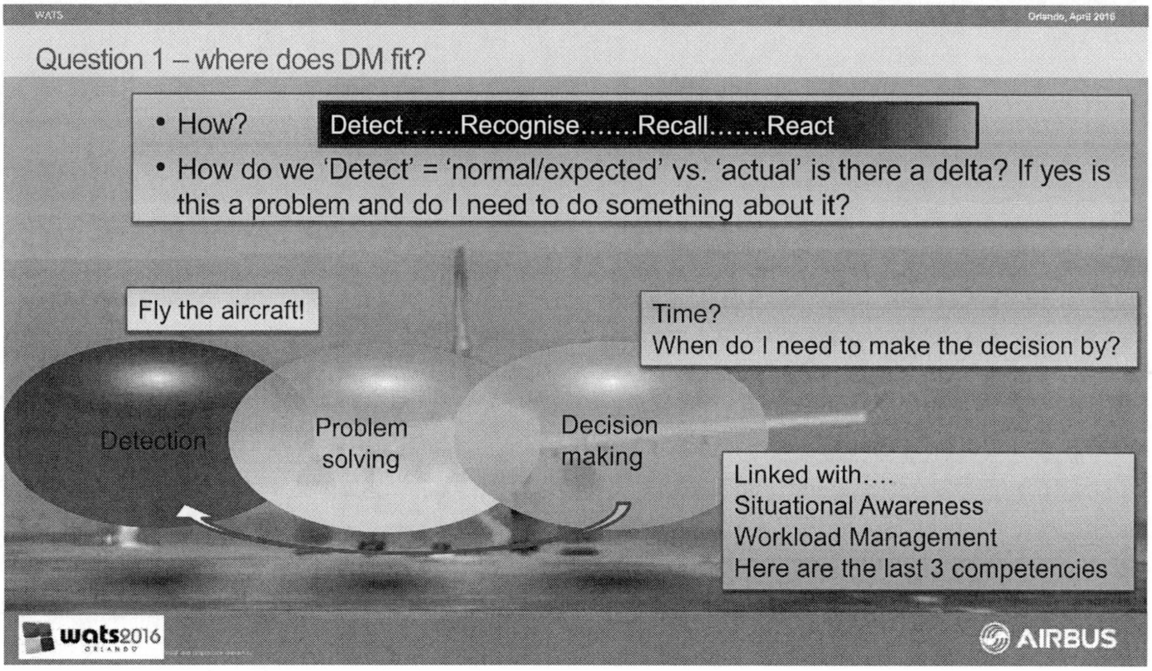
WATS
Orlando, April 2016
Question 1 – where does DM fit?
• How? Detect…….Recognise…….Recall…….React
• How do we 'Detect' = 'normal/expected' vs. 'actual' is there a delta? If yes is this a problem and do I need to do something about it?
Fly the aircraft!
Time?
When do I need to make the decision by?
Detection
Problem solving
Decision making
Linked with….
Situational Awareness
Workload Management
Here are the last 3 competencies
wats2016 ORLANDO
AIRBUS

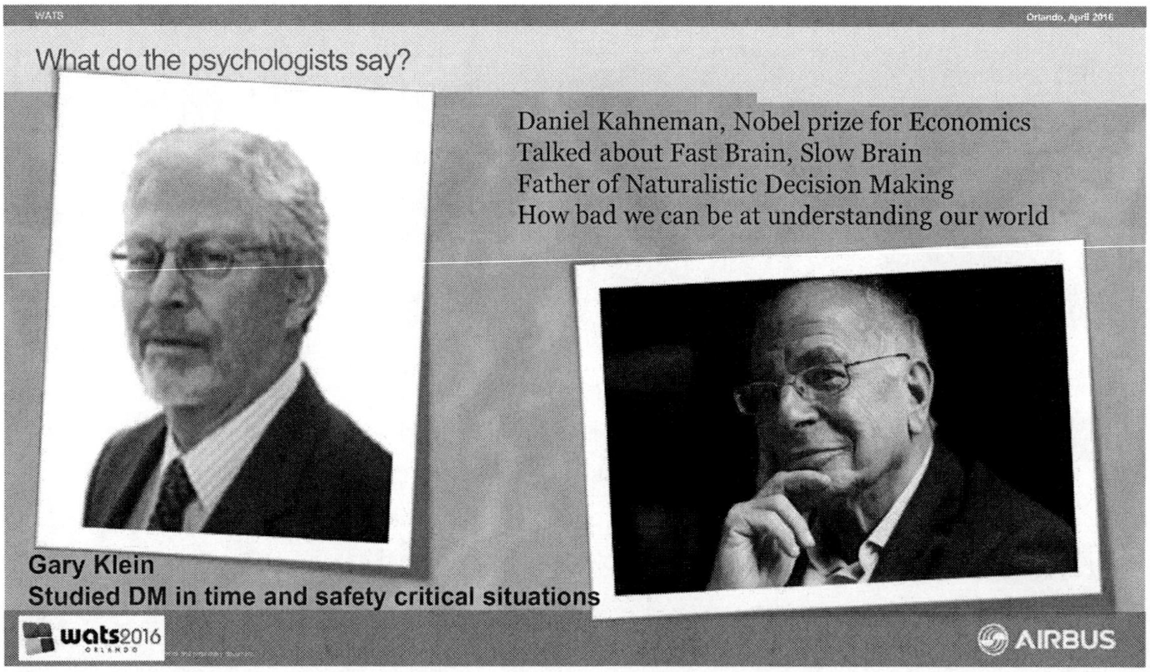
WATS
Orlando, April 2016
What do the psychologists say?
Daniel Kahneman, Nobel prize for Economics
Talked about Fast Brain, Slow Brain
Father of Naturalistic Decision Making
How bad we can be at understanding our world
Gary Klein
Studied DM in time and safety critical situations
wats2016 ORLANDO
AIRBUS

WATS Orlando, April 2016

Question 2 – it's important that pilots make good decisions but what is a good decision?

wats2016 ORLANDO AIRBUS

WATS Orlando, April 2016

What methods can we use? 8 DM methods

"If you are flying and you get a feeling that something isn't right it's probably because ***something isn't right***"

Name	Resources?	Useful to crew?
Trial and Error	Takes time!	No (training)
Analytic	Data and Time	No (time?)
Balance Sheet	Time	No (maybe?)
Flipism	Coin?	No!
Repeat	Experience	Maybe
Satisficing	Reduced	Yes!
Pattern Matching	Experience	Yes
Intuition (gut feel)	Experience	Yes/No!

Satisficing – Herbert Simon, 1978, Nobel Prize for Economics

AIRBUS

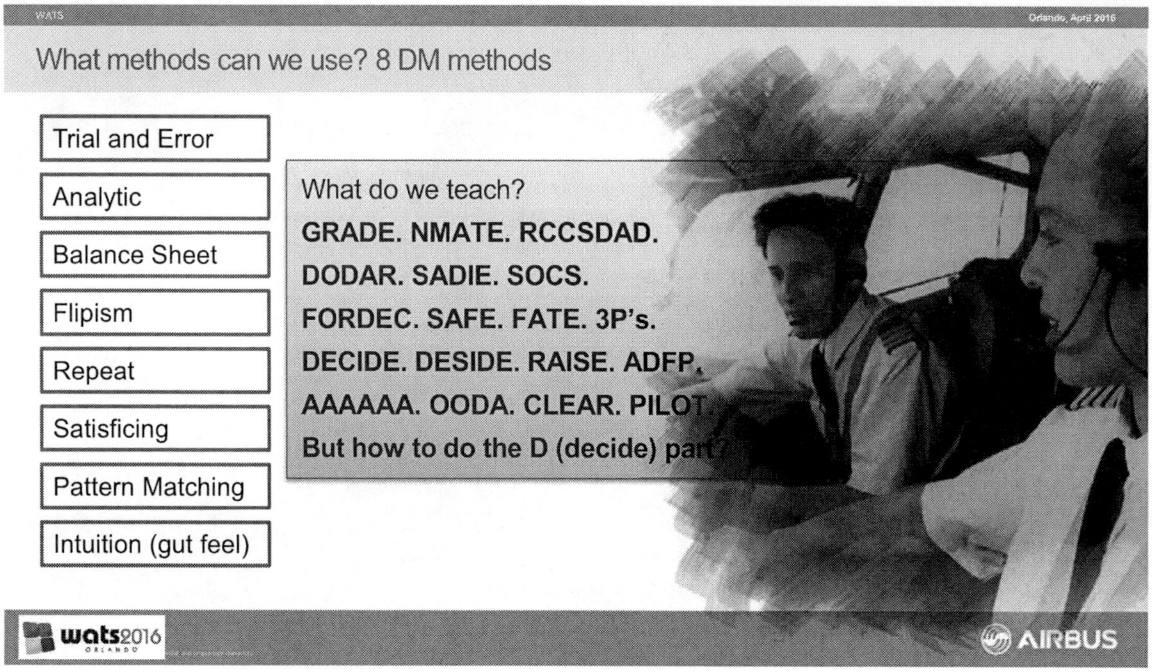
WATS
Orlando, April 2016
What methods can we use? 8 DM methods
Trial and Error
Analytic
Balance Sheet
Flipism
Repeat
Satisficing
Pattern Matching
Intuition (gut feel)
What do we teach?
GRADE. NMATE. RCCSDAD.
DODAR. SADIE. SOCS.
FORDEC. SAFE. FATE. 3P's.
DECIDE. DESIDE. RAISE. ADFP.
AAAAAA. OODA. CLEAR. PILOT.
But how to do the D (decide) part?
wats2016
ORLANDO
AIRBUS

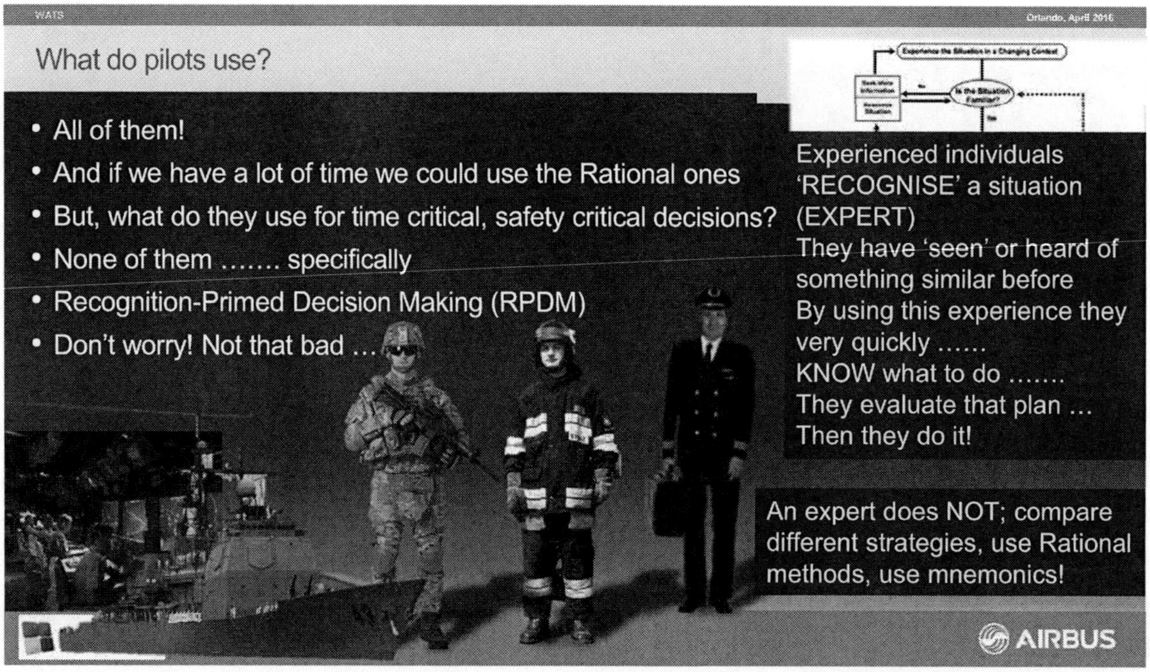
WATS
Orlando, April 2016
What do pilots use?
• All of them!
• And if we have a lot of time we could use the Rational ones
• But, what do they use for time critical, safety critical decisions?
• None of them .... specifically
• Recognition-Primed Decision Making (RPDM)
• Don't worry! Not that bad ...
Experienced individuals 'RECOGNISE' a situation (EXPERT)
They have 'seen' or heard of something similar before
By using this experience they very quickly ......
KNOW what to do ....
They evaluate that plan ...
Then they do it!
An expert does NOT; compare different strategies, use Rational methods, use mnemonics!
AIRBUS

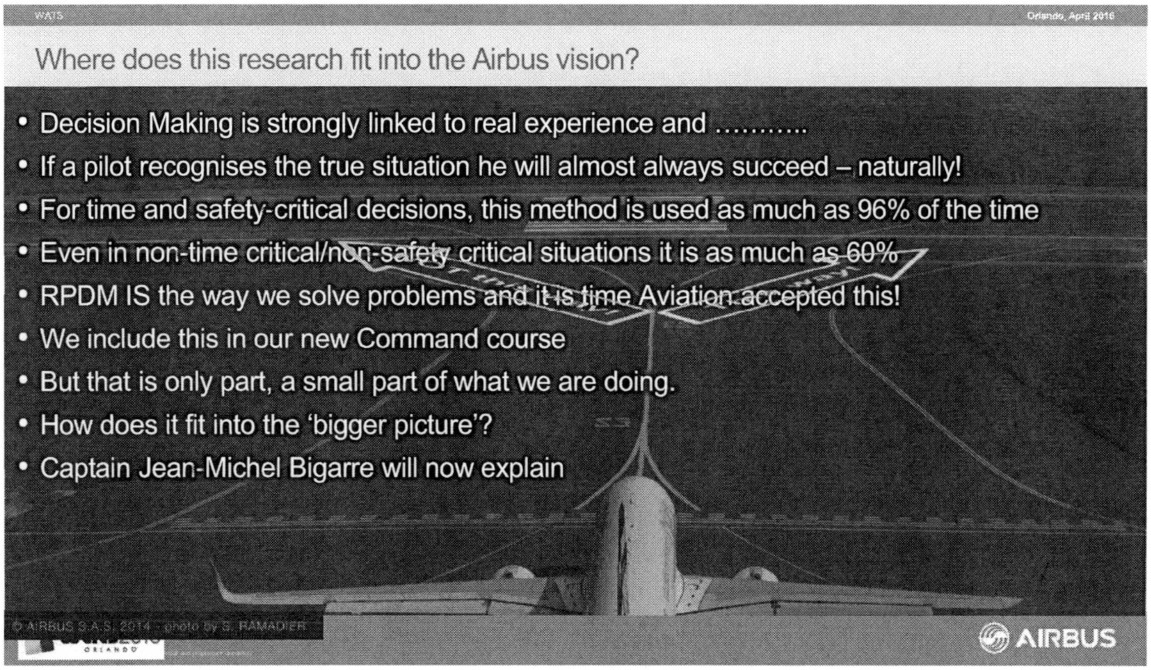
WATS
Orlando, April 2016
Where does this research fit into the Airbus vision?
• Decision Making is strongly linked to real experience and ………..
• If a pilot recognises the true situation he will almost always succeed – naturally!
• For time and safety-critical decisions, this method is used as much as 96% of the time
• Even in non-time critical/non-safety critical situations it is as much as 60%
• RPDM IS the way we solve problems and it is time Aviation accepted this!
• We include this in our new Command course
• But that is only part, a small part of what we are doing.
• How does it fit into the 'bigger picture'?
• Captain Jean-Michel Bigarre will now explain
© AIRBUS S.A.S. 2014 - photo by S. RAMADIER
AIRBUS

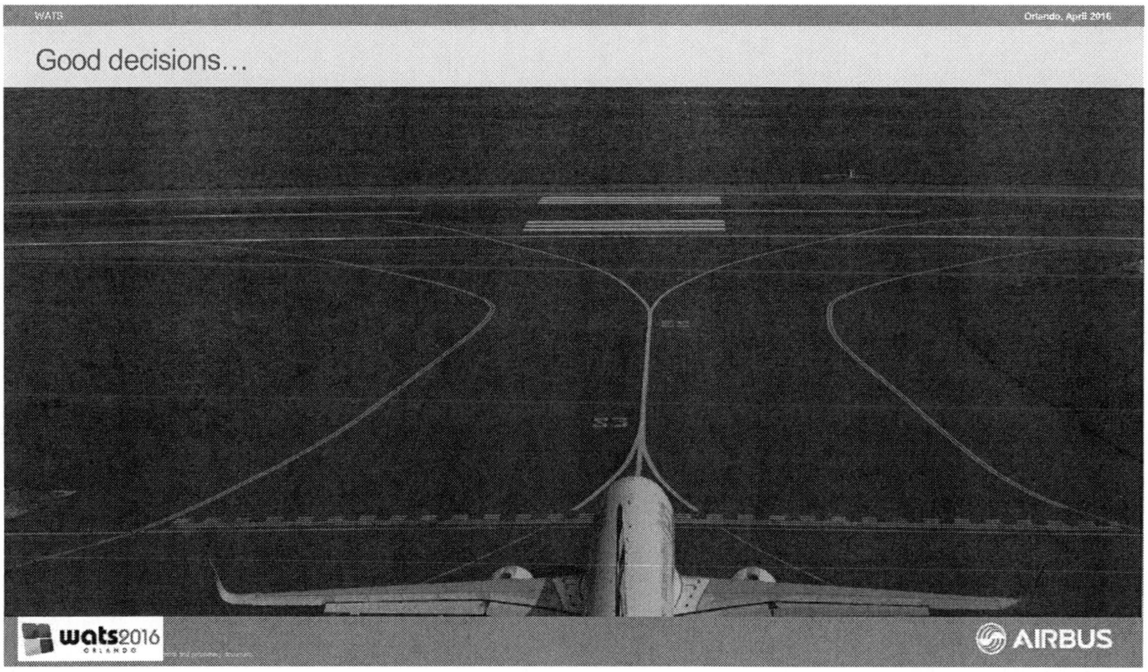
WATS
Orlando, April 2016
Good decisions…
wats2016
ORLANDO
AIRBUS

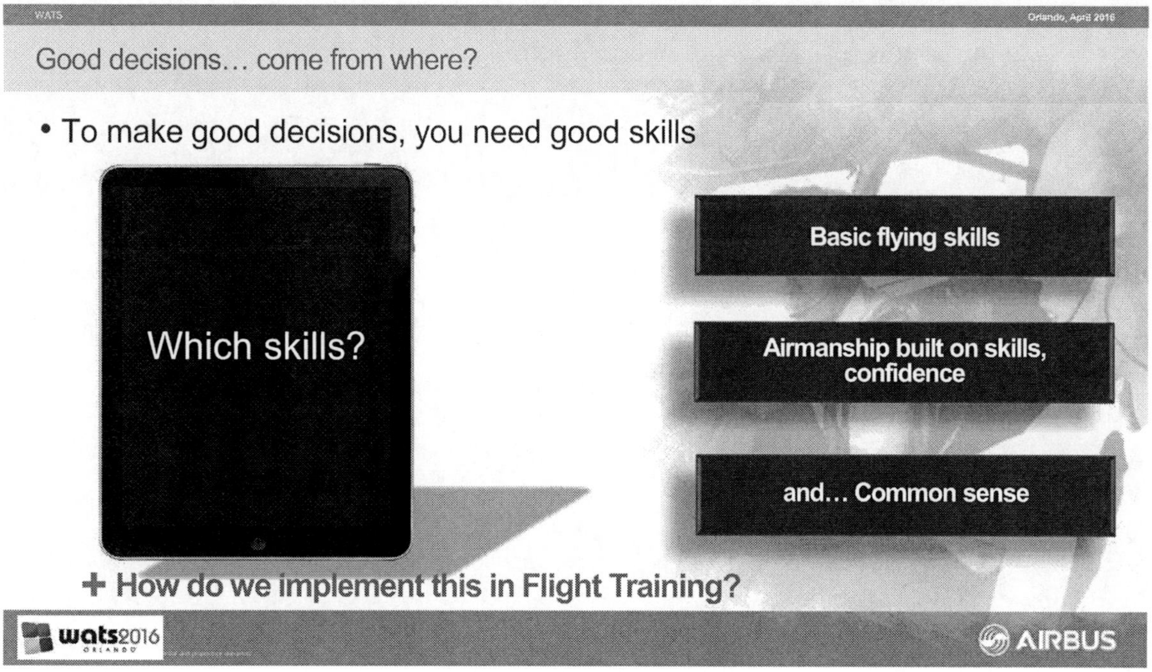
WATS
Orlando, April 2016
Good decisions… come from where?
• To make good decisions, you need good skills
Which skills?
Basic flying skills
Airmanship built on skills, confidence
and… Common sense
+ How do we implement this in Flight Training?
wats2016 ORLANDO
AIRBUS

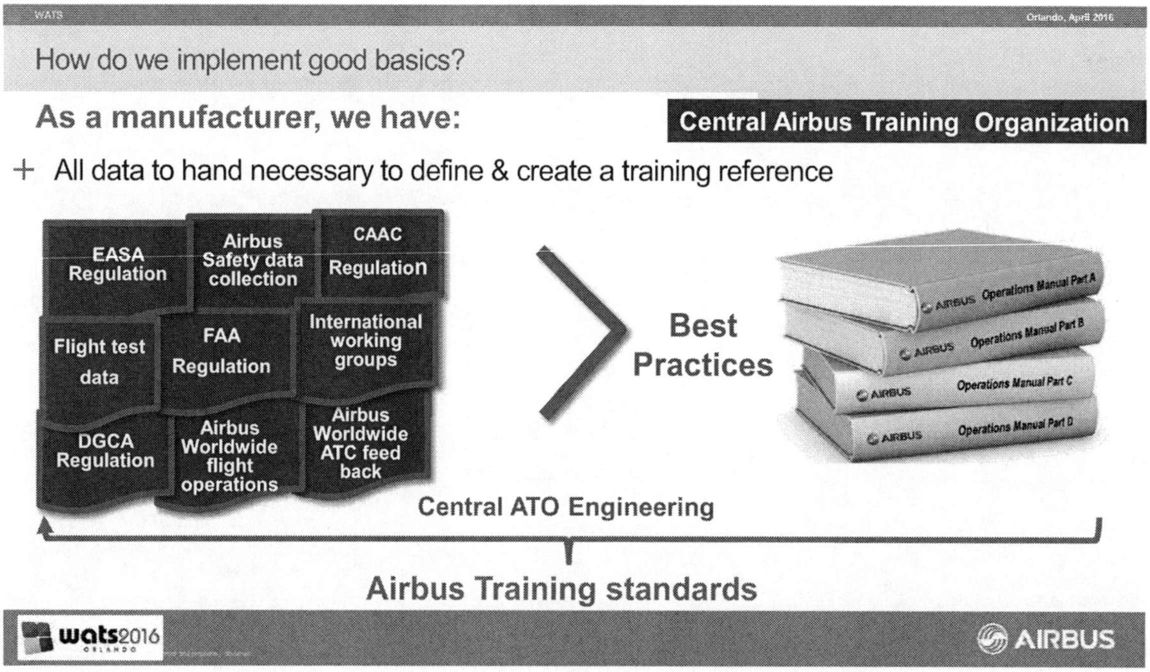
WATS
Orlando, April 2016
How do we implement good basics?
As a manufacturer, we have:
Central Airbus Training Organization
+ All data to hand necessary to define & create a training reference
EASA Regulation
Airbus Safety data collection
CAAC Regulation
Flight test data
FAA Regulation
International working groups
DGCA Regulation
Airbus Worldwide flight operations
Airbus Worldwide ATC feed back
Best Practices
AIRBUS Operations Manual Part A
AIRBUS Operations Manual Part B
AIRBUS Operations Manual Part C
AIRBUS Operations Manual Part D
Central ATO Engineering
Airbus Training standards
wats2016 ORLANDO
AIRBUS

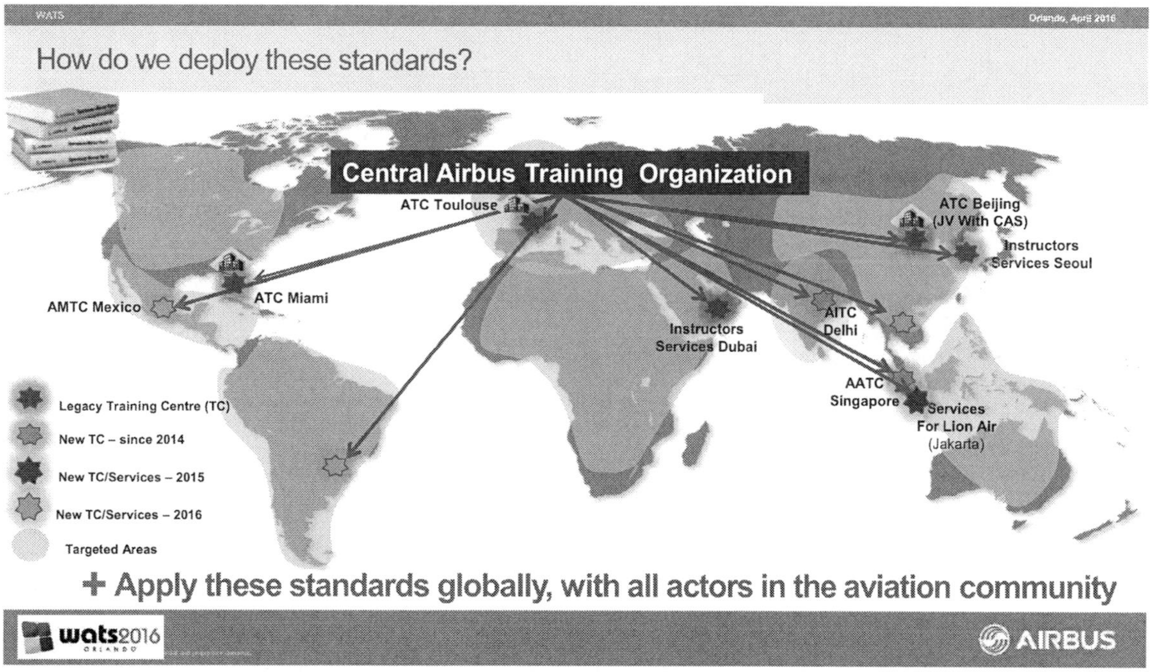
WATS
Orlando, April 2016
How do we deploy these standards?
Central Airbus Training Organization
ATC Toulouse
ATC Beijing (JV With CAS)
Instructors Services Seoul
ATC Miami
AMTC Mexico
Instructors Services Dubai
AITC Delhi
AATC Singapore
Services For Lion Air (Jakarta)
Legacy Training Centre (TC)
New TC – since 2014
New TC/Services – 2015
New TC/Services – 2016
Targeted Areas
+ Apply these standards globally, with all actors in the aviation community
wats2016 ORLANDO
AIRBUS

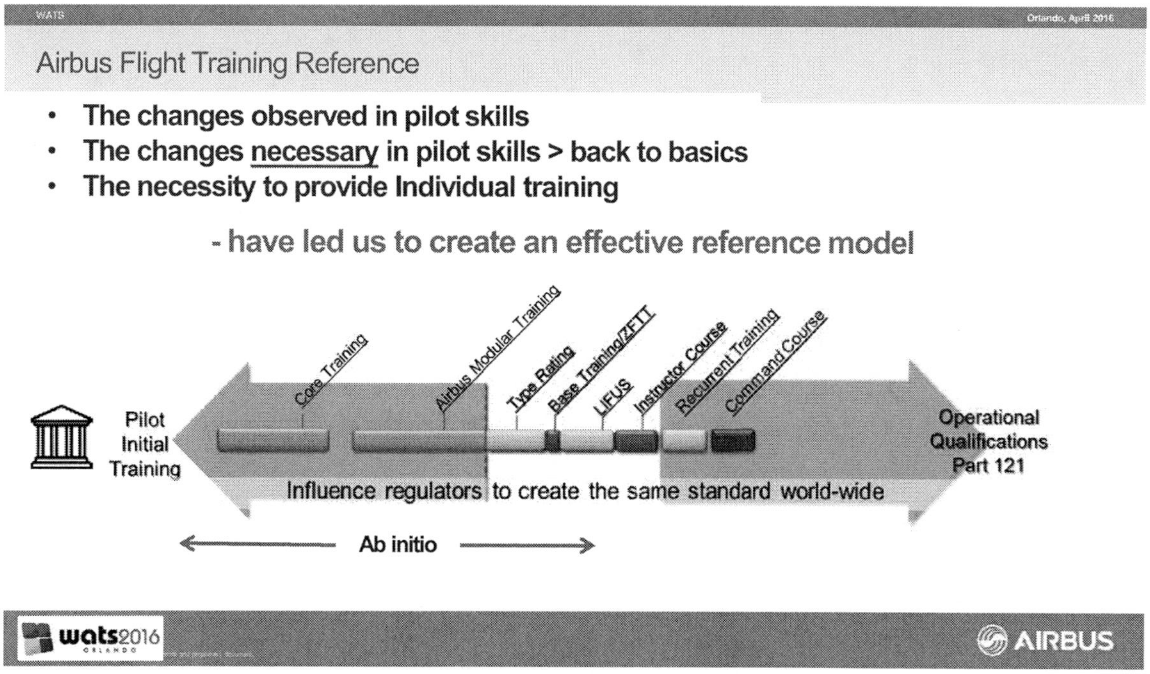
WATS
Orlando, April 2016
Airbus Flight Training Reference
• The changes observed in pilot skills
• The changes necessary in pilot skills > back to basics
• The necessity to provide Individual training
- have led us to create an effective reference model
Core Training
Airbus Modular Training
Type Rating
Base Training/ZFTT
UFUS
Instructor Course
Recurrent Training
Command Course
Pilot Initial Training
Operational Qualifications Part 121
Influence regulators to create the same standard world-wide
Ab initio
wats2016 ORLANDO
AIRBUS

WATS Orlando, April 2016

What Needs to Change in Training:

Educate pilots – starting from university to pilots specificities and competencies

Aviation Community to provide means to do it and maintain it

Airbus is doing its part and deploying its strategy to do this

+ Make sure the basics are understood, integrated and remain resilient

wats2016 ORLANDO AIRBUS

WATS Orlando, April 2016

Any Questions …

Thank you

A mutual interest to have airlines flying safely with optimized training

wats2016 ORLANDO AIRBUS

WATS

Orlando, April 2016

wats2016 ORLANDO

AIRBUS

CLASSROOMS
25 - 27
NEXT GENERATION PILOT TRAINING NEEDS
WATS 2016
CAE

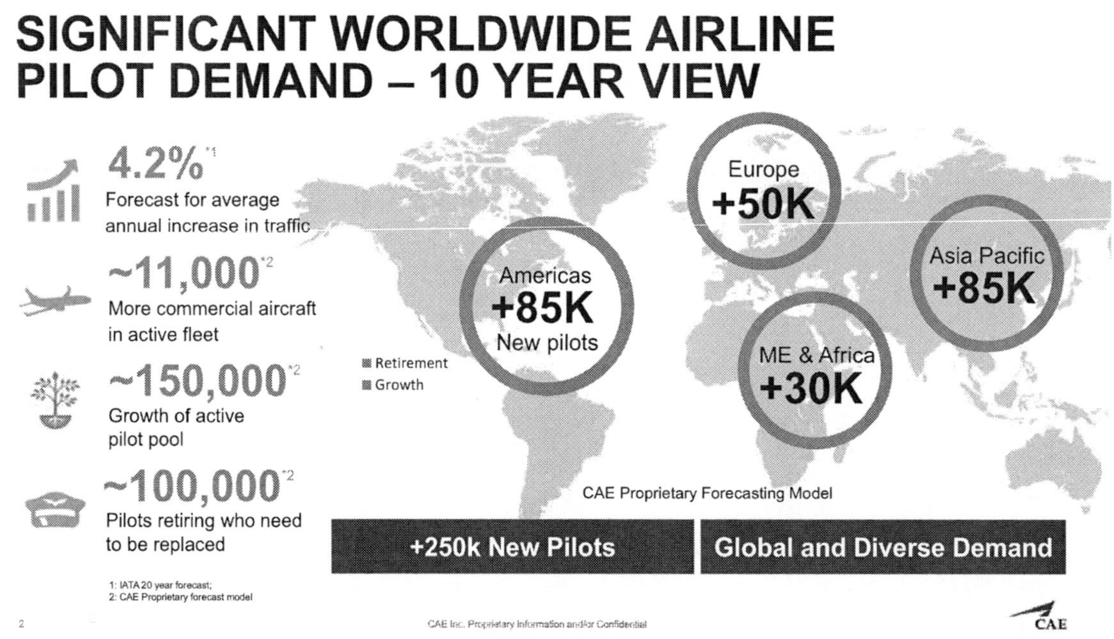
SIGNIFICANT WORLDWIDE AIRLINE PILOT DEMAND – 10 YEAR VIEW
4.2%*1
Forecast for average annual increase in traffic
~11,000*2
More commercial aircraft in active fleet
~150,000*2
Growth of active pilot pool
~100,000*2
Pilots retiring who need to be replaced
1: IATA 20 year forecast;
2: CAE Proprietary forecast model
Europe +50K
Americas +85K New pilots
Asia Pacific +85K
ME & Africa +30K
Retirement
Growth
CAE Proprietary Forecasting Model
+250k New Pilots
Global and Diverse Demand
2
CAE Inc. Proprietary Information and/or Confidential
CAE

AIRLINE CRITICAL TRAINING REQUIREMENTS, GLOBALLY

1,000+
Cadet pilots graduated / year

120,000+
Pilots trained / year

1,000+
Pilots on assignment

200+
Airline training leaders interviewed

CAE
Experience & Interactions as leading training partner for airlines, globally

Pilots who
Operate your way

Pilots who
Master Critical Situations

Pilots who
Develop Faster

AIRLINE CRITICAL TRAINING REQUIREMENTS, GLOBALLY

Pilots who
Operate your way

“We are growing quickly, it's a challenge for my pilots to know our Standard Procedures”

“If you have no standardization, you lose system safety”

“My new pilots are accessing material from all over the internet, they forget what SOP is, and sometimes this is even true for my instructors”

AIRLINE CRITICAL TRAINING REQUIREMENTS, GLOBALLY

Pilots who

Master Critical Situations

“I am receiving 2 aircrafts every month, I need my pilots to be ready”

“My pilots need to be proficient in situations they rarely experience”

“I want to be absolutely sure my pilots are compliant to the regulatory safety requirements”

5 CAE Inc. Proprietary Information and/or Confidential CAE

AIRLINE CRITICAL TRAINING REQUIREMENTS, GLOBALLY

Pilots who

Develop Faster

“I want to be absolutely sure my pilots can make the right decisions”

“We are going through a generation shift...”

“We need a system that trains pilots by competencies, not by hours”

6 CAE Inc. Proprietary Information and/or Confidential CAE

REAL LIFE EXAMPLES OF CAE'S AIRLINE TRAINING PARTNERS

- Trained to airlines SOPs from day 1
- Train to their SOPs in 7 different training locations
- Extensive focus on landing proficiency
- Faster building of skills and capabilities

IndiGo
easyJet
TURKISH AIRLINES
jetBlue
JAPAN AIRLINES

SYSTEMATICALLY ADDRESS CRITICAL TRAINING REQUIREMENTS

Pilots who **Operate your way**

Pilots who **Master Critical Situations**

Pilots who **Develop Faster**

- Instructor Quality
- Adapted Courseware
- Objectively Assess Pilot
- Augmented Training Environment

HOW READY ARE YOU?

Instructor Quality	Adapted Courseware	Objectively Assess Pilot	Augmented Training Environment
☑ Instructors trained to teach ☑ Adaptability to learning style ☑ Train to SOP, by the book, from day 1	☑ Courseware assessment post SOP changes ☑ Emphasize SOPs ☑ Include CRM	☑ Identify pilot strengths/gaps ☑ Track progress	☑ Collect objective training session data ☑ Training tools to support instruction

9

CAE

CAE

THANK YOU

Captain Jon Tovani

Managing Director – Flight Training

Making a Safe System Safer

- Excellent safety record, getting better all the time
- Congressional, FAA, and NTSB support in this endeavor
- Tough to maintain this trend … how can we do this?

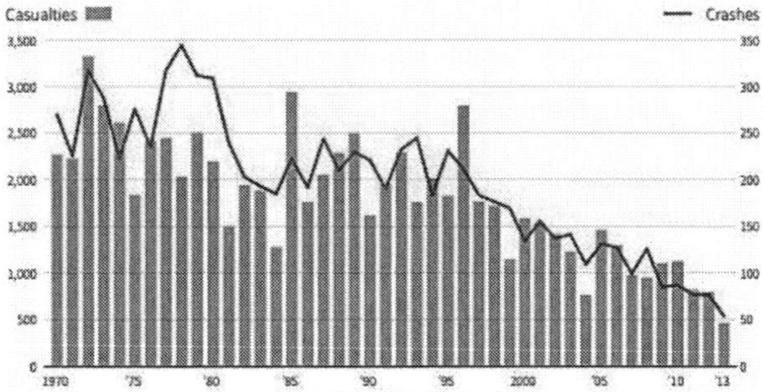

Training Manager's Focus

- The training program development processes
- Determining the high impact KSAs for pilots
- Corporate Culture & Policies and the ramifications for training programs

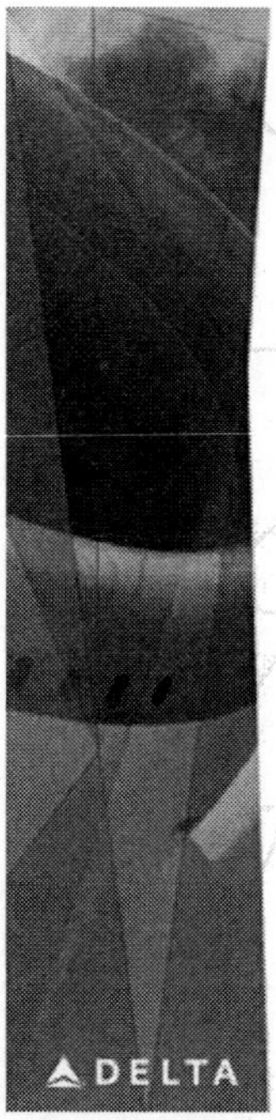

Development Process

- Inputs for evaluating our training programs are broadening
- Air Carrier Data – AQP, Safety and Ops
- Regulatory requirements
- The challenge to fit it all within the allocated training footprint

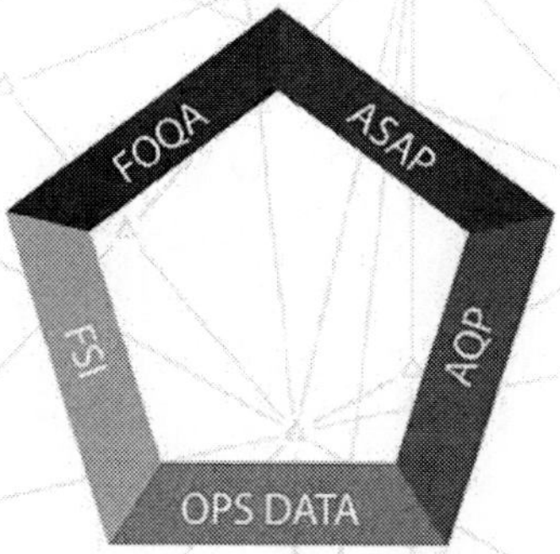

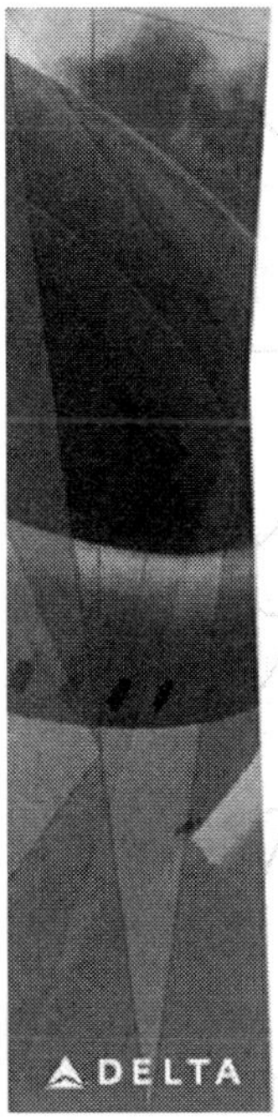

Development Process

- Aspects of the environment are changing
  - New talent – Recruiting, selection, mentoring
  - Technology – Autoflight and the dependencies is creates
  - NAS – Procedures are changing to address capacity needs

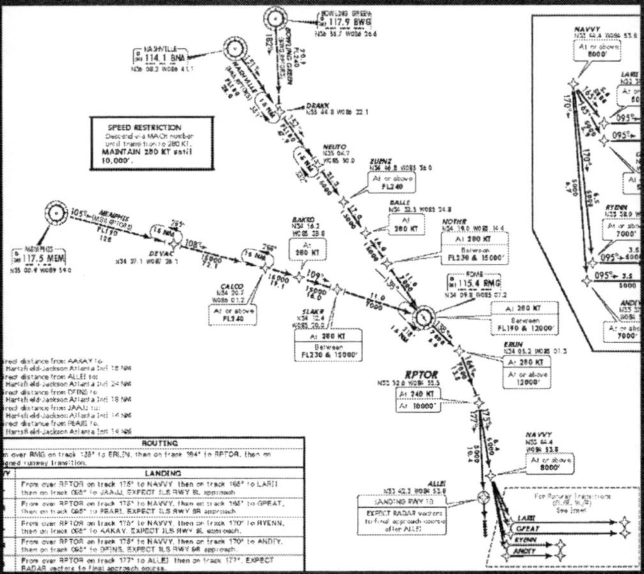

Development Process

- Proficiency – graduates capable of handling everything
  - Can't teach and practice all task and condition combinations
  - Strategy: Group KSAs of the most challenging tasks for pilots and address the underlying enabling elements, which are…

DELTA

High Impact KSA #1

FLYING!

More specifically

Managing the flight path

DELTA

Flight Path Management

- Manual flying skills
  - Learned first, easiest, most fun
  - Requires regular practice for PF and PM
  - Company policies can help or hinder

Flight Path Management

- Autoflight skills
  - Some systems are so good they breed complacency
  - More challenging with new ATC procedures
  - Instructor training required

High Impact KSA #2

CRM/TEM

Effective Monitoring

The objective is to have sufficient awareness of a situation so as to make appropriate decisions about maintaining the flight path

DELTA

CRM/TEM

Engage the team and invite input

- Plan and communicate (brief) the plan, especially if it may involve infrequently performed tasks (go-around)
- These skills can and should be applied during preflight phase
  - Crew briefings including cabin crew
  - Dispatch readiness
    - MEL assessment

Leadership is required

DELTA

High Impact KSA #3

Procedural Execution

Procedural Execution

- Provide review via periodic distributed training
  - Selected based on airline safety/operational data
- Provide practice of a sampling during simulator training
  - Avoid predictable routines
  - There's value in surprise
- Refresh during crew briefings
  - "In the event we around, I'll select TOGA, you confirm SRS…

Corporate Culture & Policies

- Safety first from the top down
  - Above financial and operational results, of course!
- Robust involvement in safety programs
  - AQP, FOQA, ASAP, SMS
  - Reporting culture
  - Compliancy philosophy
- Commitment to continuous improvement
  - Data driven program evaluation

DELTA

In Summary

Training Manager's levers to pull

- The training program development processes
  - Inputs, Environment, Proficiency
- Provide training in the high impact KSAs
  - Flight Path, CRM & Monitoring, Procedures
- Shape corporate culture & policies
  - Safety, Compliance, Reporting

This is how we make a safe system, even safer!

DELTA

Thank You

A Whole New (Mager) World
Southwest Airlines University

Dan Duren

Courtesy of HGTV

Instructional Methodology
PADDIE

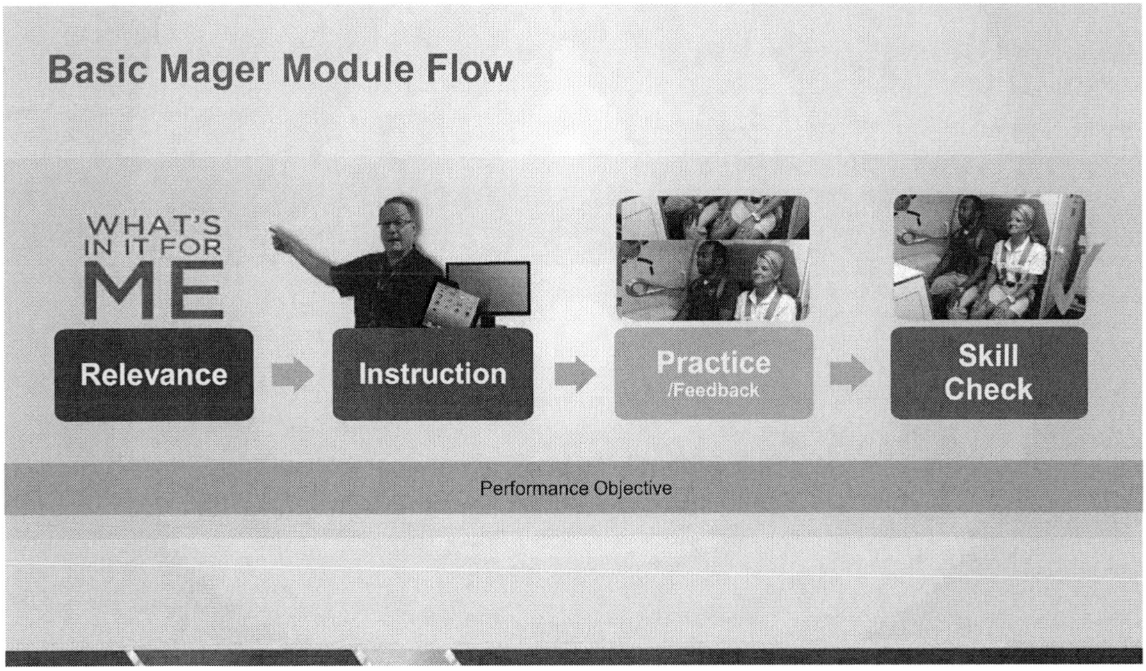
Basic Mager Module Flow
WHAT'S IN IT FOR ME
Relevance
Instruction
Practice /Feedback
Skill Check
Performance Objective

Developer Certification

CERTIFIED
Instructional Technologist

Ten day certification including demonstrating skill checks for each of the 44 skills mastered in the courses:

- Criterion-Referenced Instruction (CRI)
- Instructional Module Development (IMD)
- Building Better Job Aids (BBJA)

Mager Consortium
Common Sense in Workplace Performance

Current Program

- 5 week program
- Topical/modular approach
- 24 guest speakers
- Limited linkages to company orientation specific to Flight Attendant role
- 15 Instructors
- 9 knowledge exams

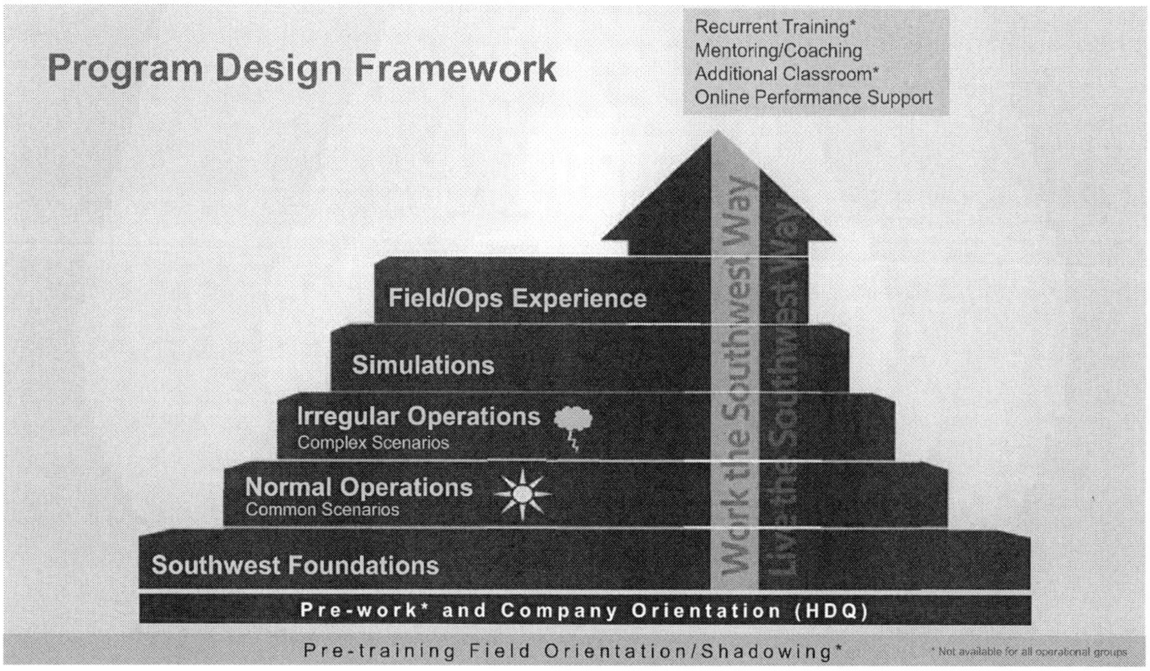
Program Design Framework
Recurrent Training*
Mentoring/Coaching
Additional Classroom*
Online Performance Support
Field/Ops Experience
Simulations
Irregular Operations
Complex Scenarios
Normal Operations
Common Scenarios
Southwest Foundations
Work the Southwest Way
Live the Southwest Way
Pre-work* and Company Orientation (HDQ)
Pre-training Field Orientation/Shadowing*
* Not available for all operational groups

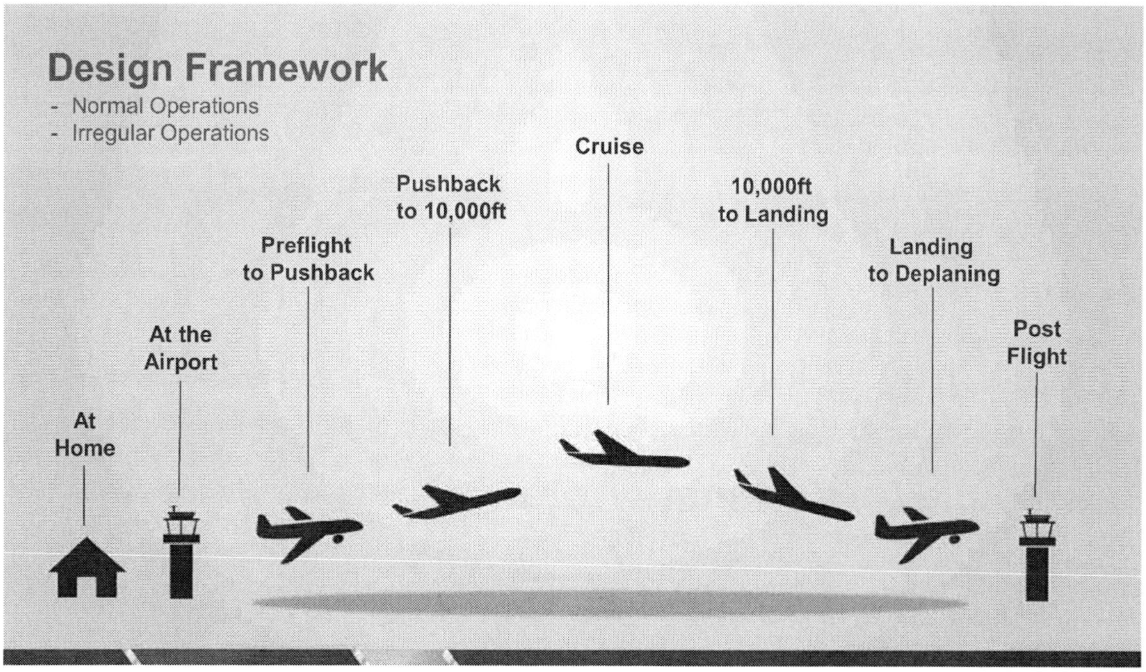
Design Framework
- Normal Operations
- Irregular Operations
At Home
At the Airport
Preflight to Pushback
Pushback to 10,000ft
Cruise
10,000ft to Landing
Landing to Deplaning
Post Flight

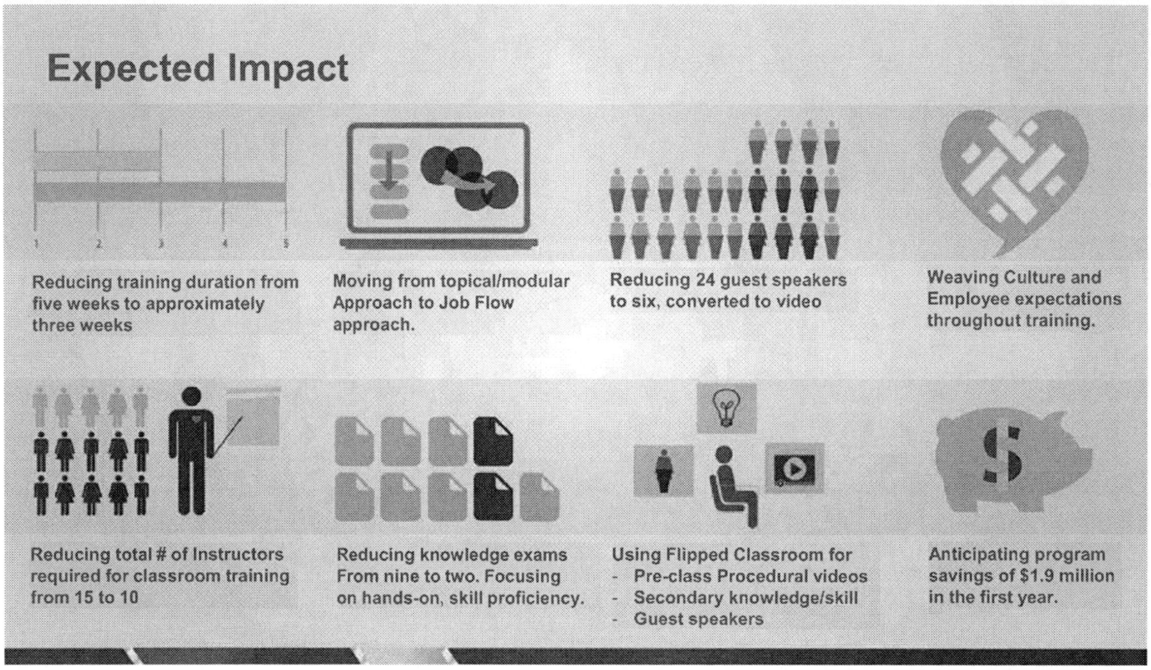
Expected Impact
Reducing training duration from five weeks to approximately three weeks
Moving from topical/modular Approach to Job Flow approach.
Reducing 24 guest speakers to six, converted to video
Weaving Culture and Employee expectations throughout training.
Reducing total # of Instructors required for classroom training from 15 to 10
Reducing knowledge exams From nine to two. Focusing on hands-on, skill proficiency.
Using Flipped Classroom for
- Pre-class Procedural videos
- Secondary knowledge/skill
- Guest speakers
Anticipating program savings of $1.9 million in the first year.

A Whole New (Mager) World
Southwest Airlines University
Dan Duren

Using Video to Enhance cabin crew training effectiveness & efficiency

Trevor Dale FRAeS

Retired Training Captain British Airways

www.Atrainability.co.uk

vosaic™

Atrainability
Share the Learning

Why is the Instructor there?

To take the student further faster

Atrainability
Share the Learning

What are the learning problems?

- Knowledge
- Skill
- Attitude

Technical performance is only part of the problem

Non-Technical performance is just as important

Behaviour – under stress or fear

Stress Solutions

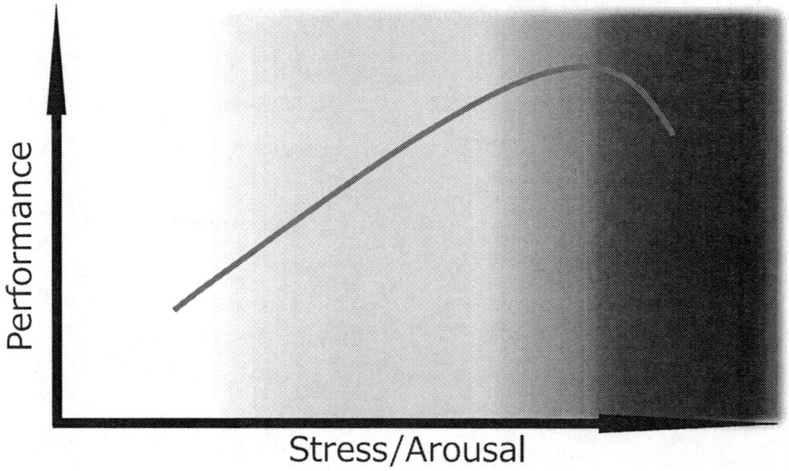

Content © atrainability 2016

Atrainability
Share the Learning

Stress Solutions

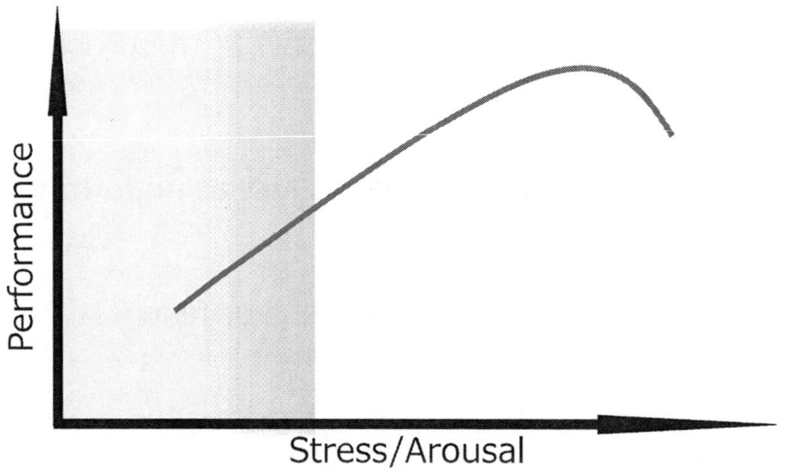

Content © atrainability 2016

Atrainability
Share the Learning

Stress Solutions

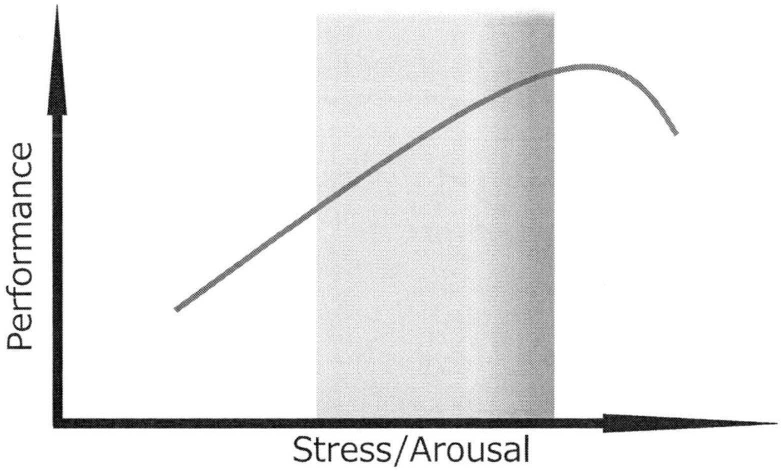

Content © atrainability 2016

Stress Solutions

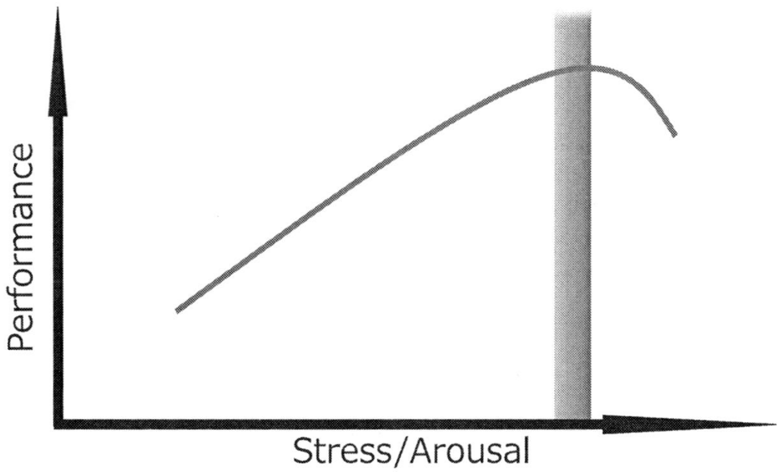

Content © atrainability 2016

Stress Solutions

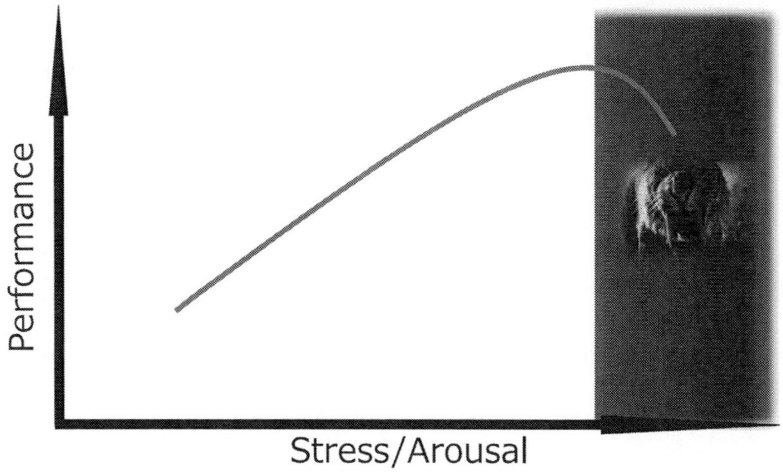

Content © atrainability 2016

Stress Solutions

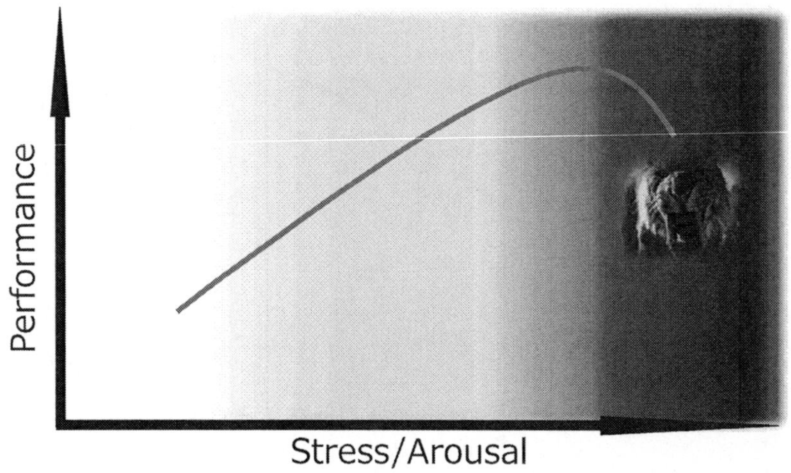

Content © atrainability 2016

Non-Normal - Evacuation

Fire-fighting

Normal - Passenger service

Dealing with toilet queues

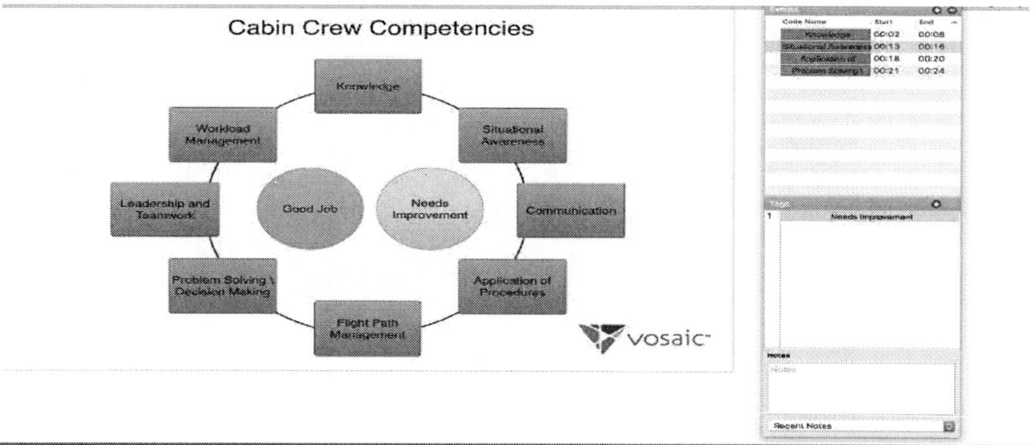

Las Vegas September 2015

Content © atrainability 2016

Atrainability
Share the Learning

Las Vegas September 2015

Content © atrainability 2016

Atrainability
Share the Learning

Toronto August 2005

Atrainability
Share the Learning

vosaic™

Evacuation drills

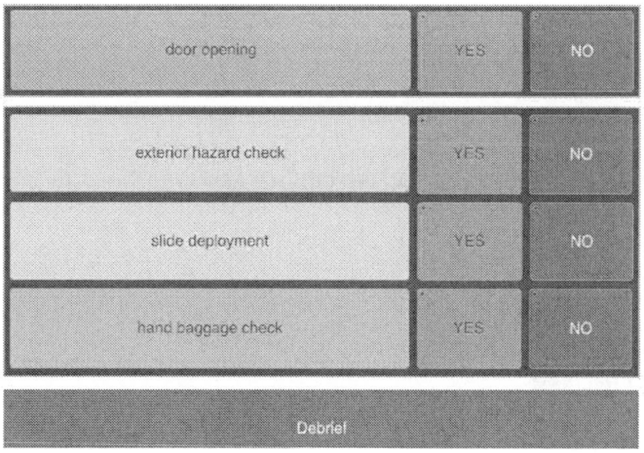

Observer

Atrainability
Share the Learning

Trainer Aid:

- Video capture and mark the timeline for easy playback

Training Management Aid:

- Video capture and mark the timeline for easy playback
- Capture the data as .xml or .csv
- Use the data to compare
  - Individuals
  - Instructors
  - courses

Instructor iPad with customised buttons for

- Technical performance
- Non-technical performance
- Anything you wish

How adults learn

Tell me and I will forget

Show me and I will remember

Involve me and I will learn

Benjamin Franklin

How many Trainers does it take to change a light bulb?

Only 1 but the light-bulb must want to change

Nothing is as powerful as seeing yourself on video

Thank you

Trevor Dale FRAeS

Atrainability Ltd

Airbus Cabin Crew Training
4/15/2016
Airbus Cabin Crew Training
Claire Coleman
The transformation from classroom to virtual reality
AIRBUS

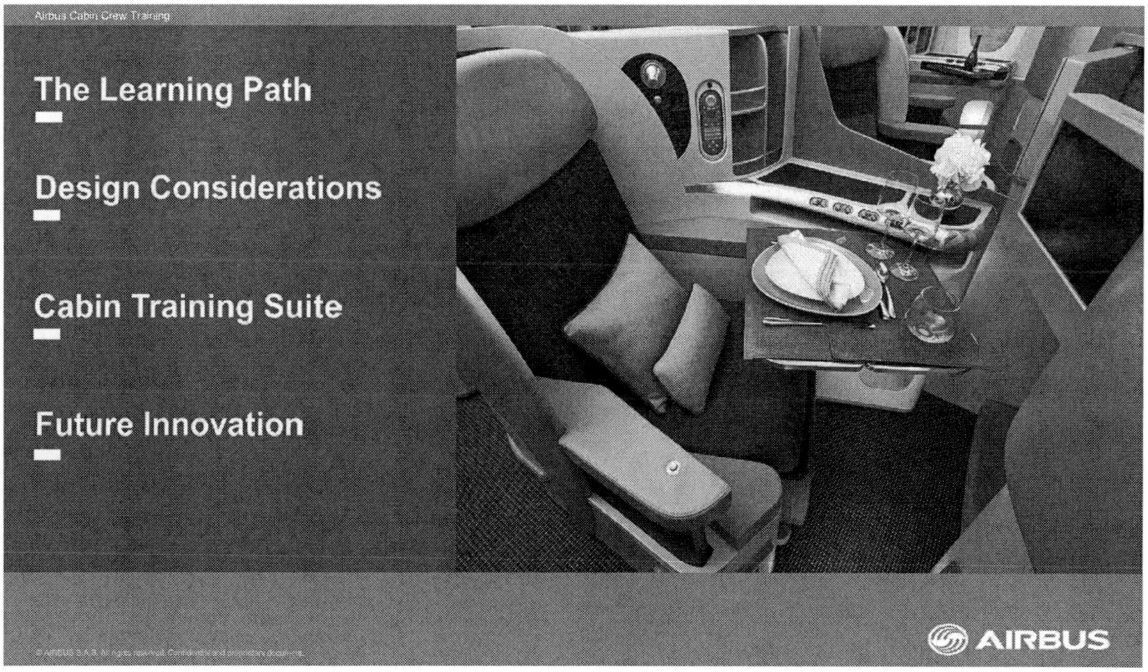
Airbus Cabin Crew Training
The Learning Path
Design Considerations
Cabin Training Suite
Future Innovation
AIRBUS

The Learning Path

Design Considerations

Cabin Training Suite

Future Innovation

AIRBUS

The Learning Path

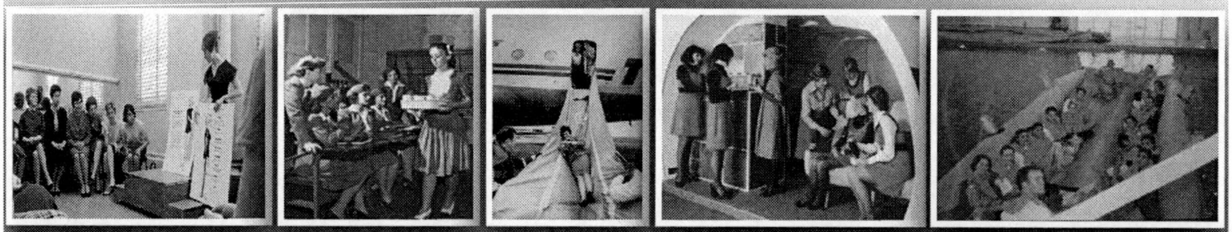

AIRBUS

Course Objectives – Airbus vs Airline

Airbus	Airline
➢ Aircraft Type Specific + detailed	➢ Aircraft Type Specific + regulations, general safety and SOP's
➢ Non licensed course	➢ Licensed course
➢ Designed for experienced personnel	➢ Designed for a range of experience

AIRBUS

The Learning Path

Design Considerations

Cabin Training Suite

Future Innovation

AIRBUS

The Learning Path

Design Considerations

Cabin Training Suite

Future Innovation

AIRBUS

Baby Boomers (1946 – 1964)

Traits:

- Read books cover to cover

Methods:

- Overhead projectors/ videos

Preferred communication style:

- In person

1950 Courtesy of Vintag.es

AIRBUS

Generation X (1965 – 1980)

Traits:
- Use index in books to find relevant section

Methods:
- Simulation and technology

Preferred communication style:
- Direct (email/text/ in person)

Generation Y (1981 – 2000)

Traits:
- Highly visual learners

Methods:
- Technology orientated
- Prefer a multi media environment

Preferred communication style:
- Direct (email/ text)

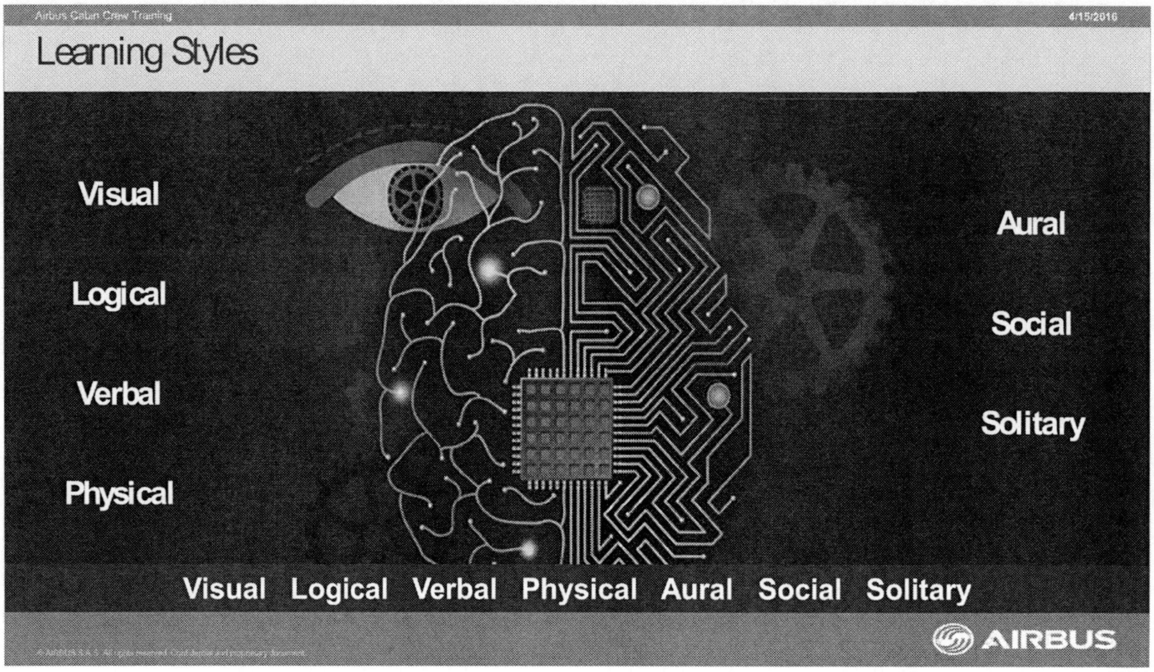
Airbus Cabin Crew Training
4/15/2016
Learning Styles
Visual
Logical
Verbal
Physical
Aural
Social
Solitary
Visual Logical Verbal Physical Aural Social Solitary
AIRBUS

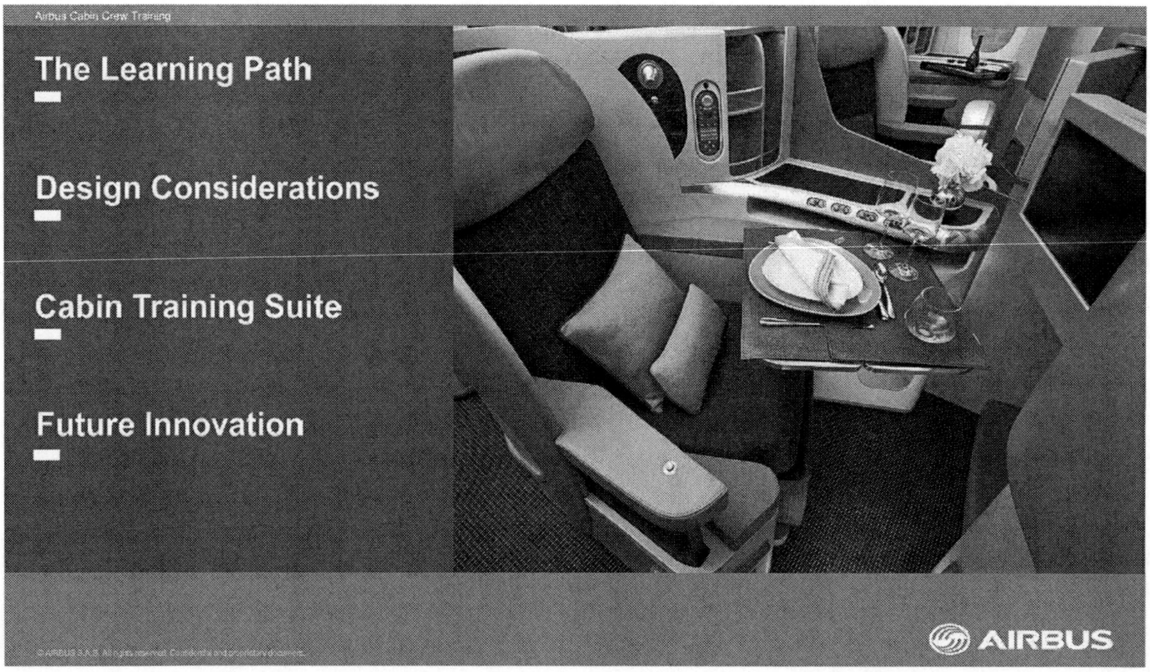
Airbus Cabin Crew Training
The Learning Path
Design Considerations
Cabin Training Suite
Future Innovation
AIRBUS

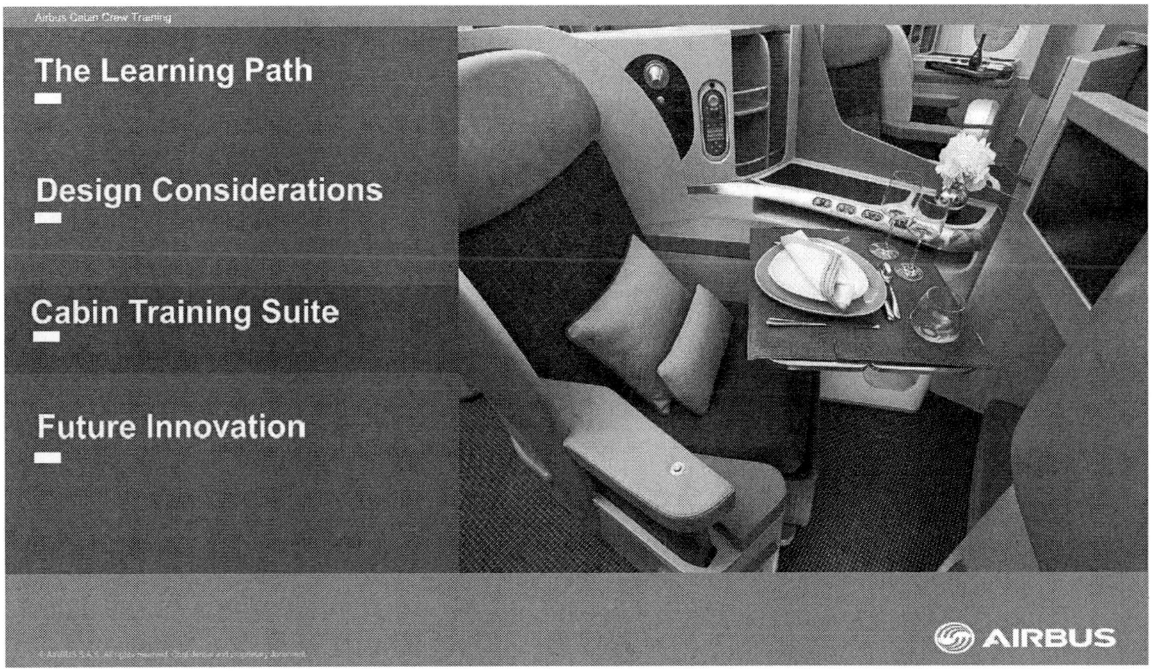
Airbus Cabin Crew Training
The Learning Path
Design Considerations
Cabin Training Suite
Future Innovation
AIRBUS

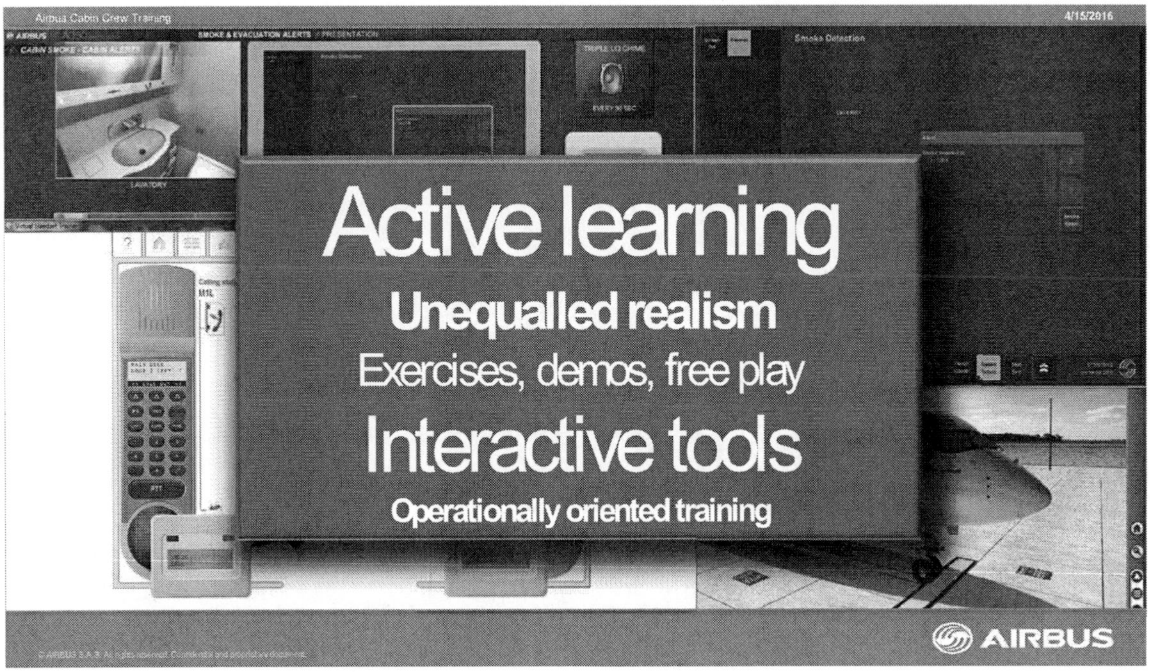
Airbus Cabin Crew Training
4/15/2016
SMOKE & EVACUATION ALERTS
Smoke Detection
LAVATORY
Active learning
Unequalled realism
Exercises, demos, free play
Interactive tools
Operationally oriented training
AIRBUS

Cabin Knowledge Modules (CKM)

Theoretical Training

Animation
Video's
Instructor led and/or
'Self Study'

Visual Aural Verbal Logical Social Solitary

AIRBUS

Virtual Cabin

Aircraft Familiarization

Aircraft walk around
Cabin familiarization
Emergency equipment location
Instructor led or
Free play

Reduce aircraft visits

Visual Verbal Physical Logical Social Solitary

AIRBUS

Virtual Handset Trainer

Handset Familiarization

Realistic training
Simulation of actions and events
Instructor led or
Free play

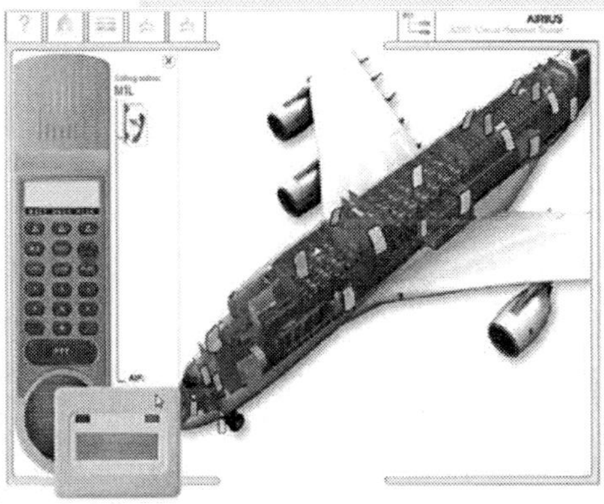

Interactive hands-on learning experience

Visual Aural Physical Logical Social Solitary

AIRBUS

Virtual Flight Attendant Panel

FAP Familiarization

Realistic training
Simulation of actions and events
Instructor led or
Free play

Saves training time on the actual aircraft

Visual Aural Verbal Physical Logical Social Solitary

AIRBUS

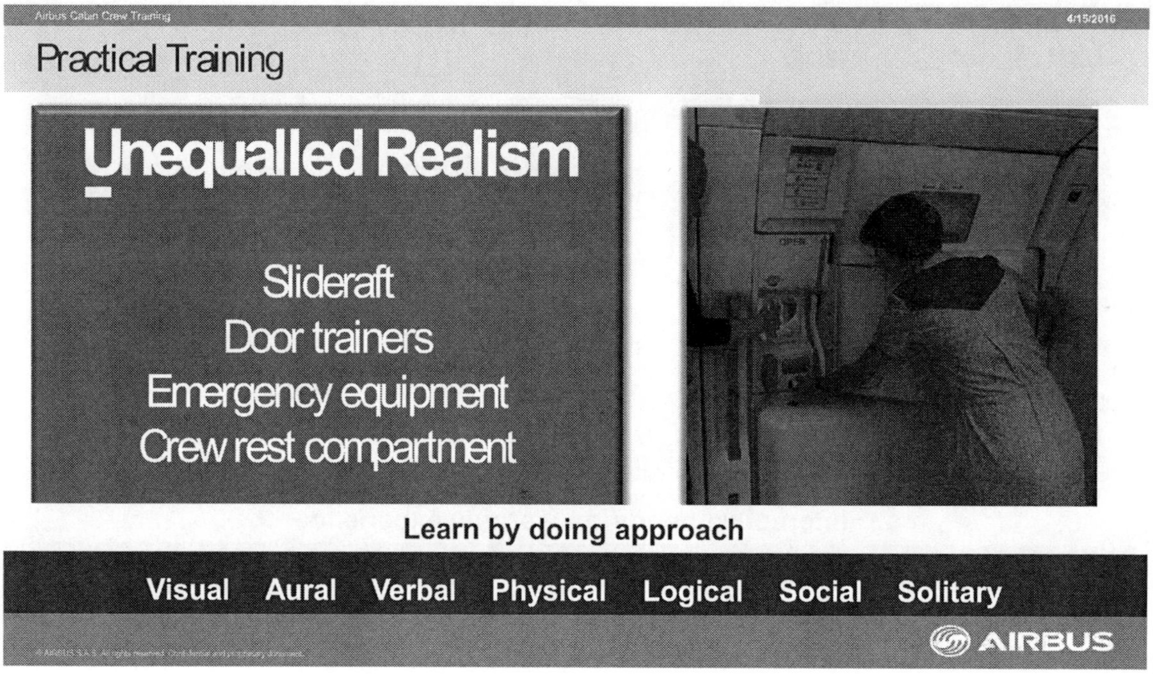
Airbus Cabin Crew Training
4/15/2016
Practical Training
Unequalled Realism
Slideraft
Door trainers
Emergency equipment
Crew rest compartment
Learn by doing approach
Visual Aural Verbal Physical Logical Social Solitary
AIRBUS

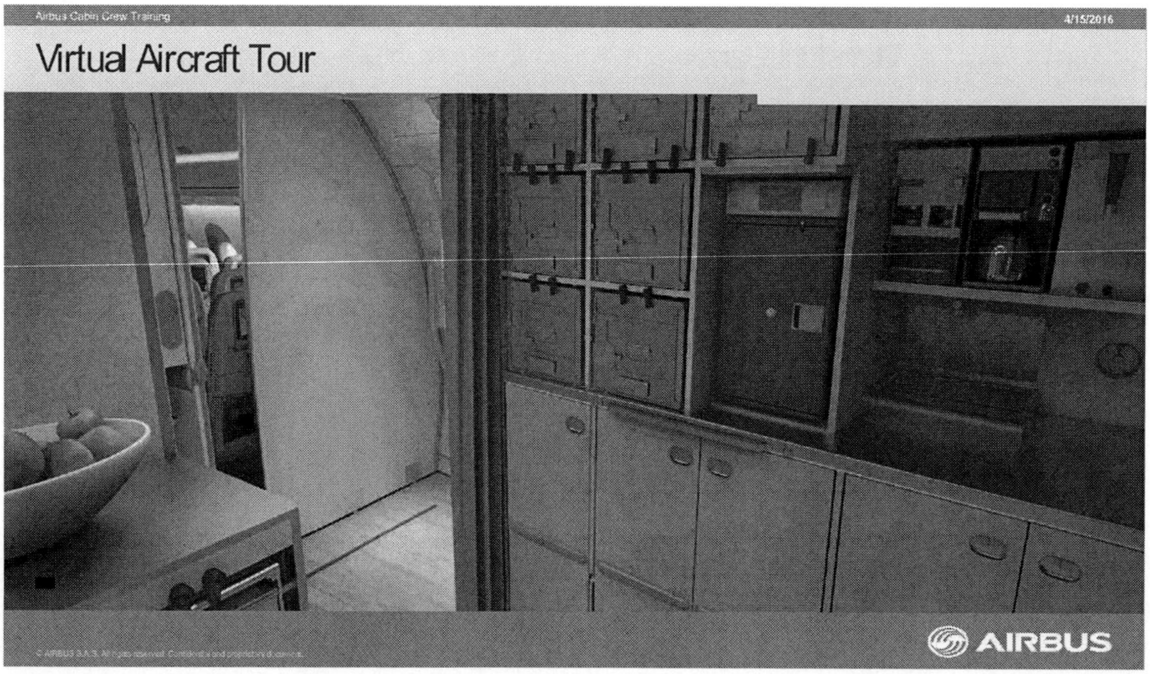
Airbus Cabin Crew Training
4/15/2016
Virtual Aircraft Tour
AIRBUS

Airbus Cabin Crew Training 4/15/2016

Virtual training in the classroom

AIRBUS

Airbus Cabin Crew Training 4/15/2016

Virtual training in the classroom

Believable

Interactive

Immersive

AIRBUS

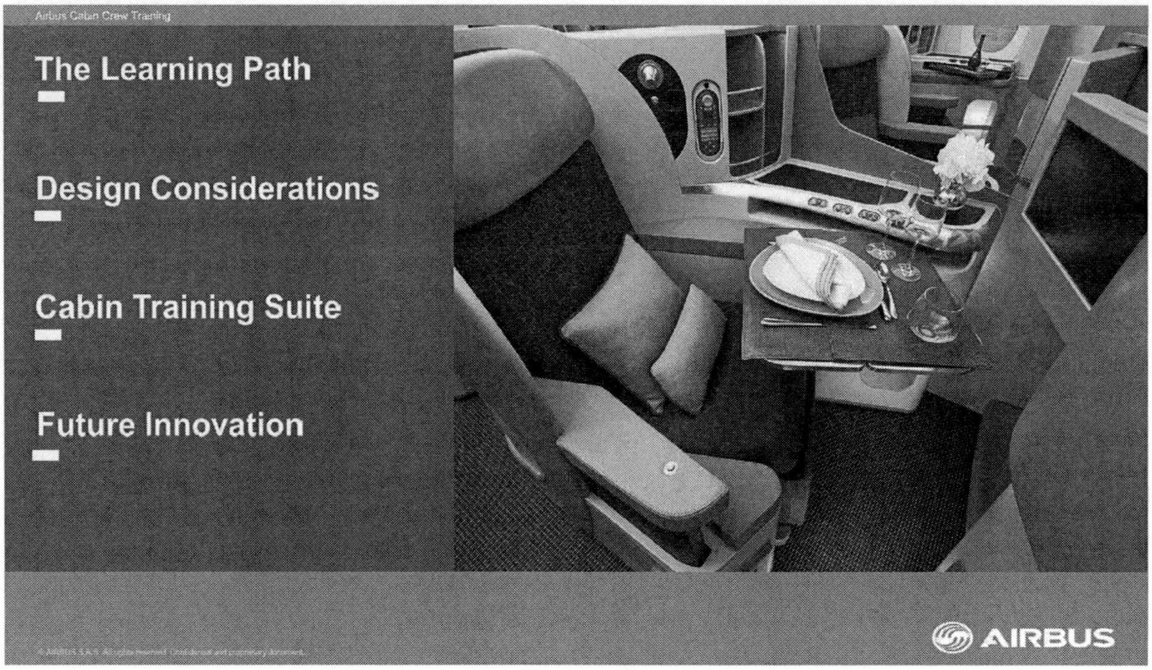
Airbus Cabin Crew Training
The Learning Path
Design Considerations
Cabin Training Suite
Future Innovation
AIRBUS

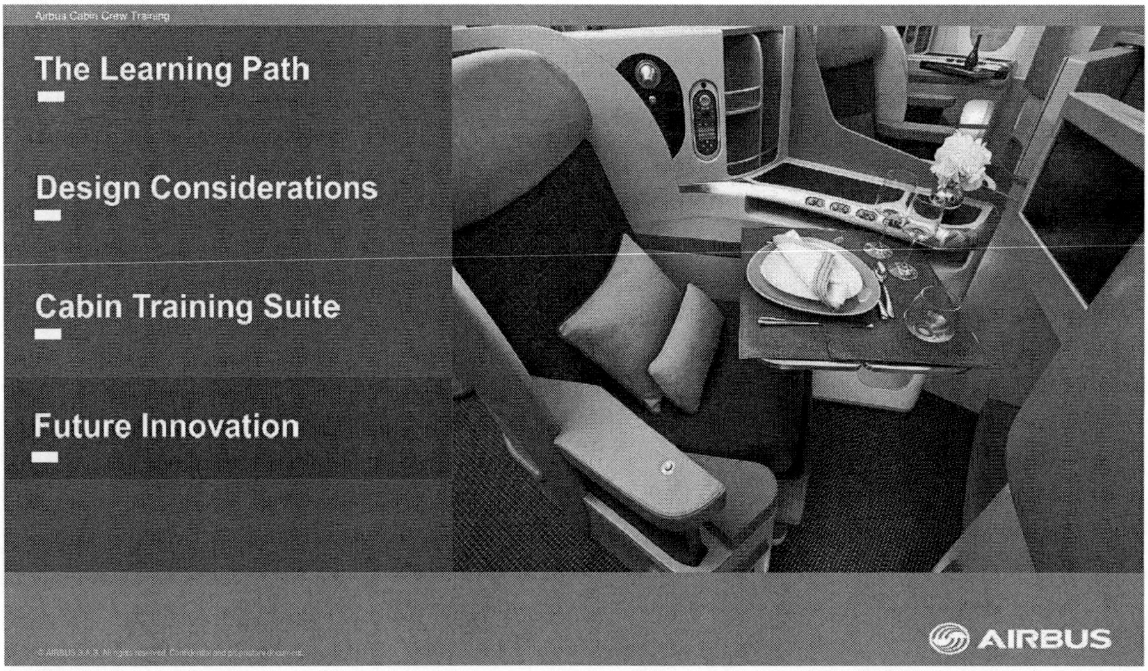
Airbus Cabin Crew Training
The Learning Path
Design Considerations
Cabin Training Suite
Future Innovation
AIRBUS

Customer Definition Centre

AIRBUS

Customer Definition Centre

AIRBUS

Customer Definition Centre

AIRBUS

Customer Definition Centre

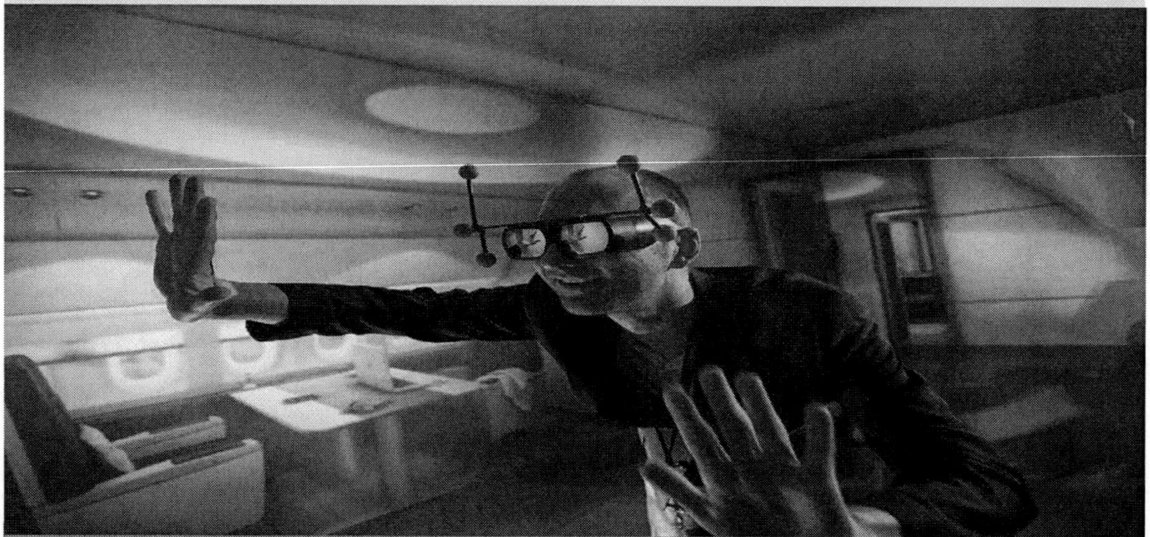

AIRBUS

Airbus Cabin Crew Training 4/15/2016

Customer Definition Centre

AIRBUS

Airbus Cabin Crew Training 4/15/2016

Customer Definition Centre

AIRBUS

Airbus Cabin Crew Training
4/15/2016
Virtual Discovery
Learn by doing
Incorporates the generation and individual learning styles
Reduce aircraft visits and cost
Active learning
Allows access to the aircraft at own pace and time
AIRBUS

Airbus Cabin Crew Training
4/15/2016
Thank You
AIRBUS

ICAO UNITING AVIATION

Fume events:
What should cabin crew know?

Martin Maurino
Safety, Efficiency and Operations Officer
International Civil Aviation Organization

WATS 2016
Orlando, 19-21 April 2015

ICAO UNITING AVIATION

Overview

- Background
- ICAO *Guidelines on Education, Training and Reporting Practices related to Fume Events* (Circular 344)
- Basic education
- Cabin crew training
- Reporting and investigation

| ICAO UNITING AVIATION

Why is this an issue?

- Various types of fumes may contaminate air supply system

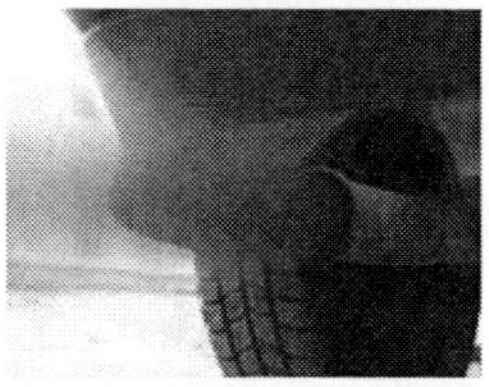

| ICAO UNITING AVIATION

Why is this an issue?

- Particular concerns on crew member exposure to oil or hydraulic fluid fumes or smoke
  - acute symptoms in flight
- Negative impact on flight safety
  - if crew are incapacitated

- Beneficial to provide guidance to crew members
  - recognize and respond to suspected air supply system-sourced fumes
  - for AMT to identify source of contaminants & correct issue

ICAO | UNITING AVIATION

Why is ICAO involved?

- Issue raised at A 38 (2013)
- Consider flight safety implications of fume events
- ITF & IFALPA requested ICAO develop guidance material
- The Technical Commission agreed
- Ad hoc working group established:

IATA ICAO IFALPA The Global Voice of Pilots ICCAIA itf

ICAO | UNITING AVIATION

ICAO Guidelines (Circular 344)

- Basic education
- Training
- Standardized reporting
- Troubleshooting
- Event investigation

ICAO UNITING AVIATION

What should basic education cover?

- Sources and types of on-board fumes
- Odour descriptors
  - to recognize presence of oil & hydraulic fluid fumes
- Potential for impairment
- Procedures to apply in fume events
- Reporting of fume events

ICAO UNITING AVIATION

Examples of potential types of fumes

Potential contaminants in the ventilation supply air	De-icing and/or anti-icing fluid Electrical faults Engine compressor wash Engine oil Exhaust (aircraft or ground vehicles) Fuel Hydraulic fluid Recirculation fan failure
Items in the cabin and/or flight deck that can be sources of fumes	Carry-on baggage Cleaning products Disinfectants Disinsectants Food items Galley equipment Lavatories

| ICAO UNITING AVIATION

Odour descriptors

- Variety of terms used to describe oil and hydraulic fluid fumes
- Odour is subjective
- Olfactory fatigue reduces ability to detect odours over time
- Education and reporting forms should include descriptors
  - typically used for air supply system-sourced fumes
- Crew should know descriptors
  - in order to effectively communicate and report fumes

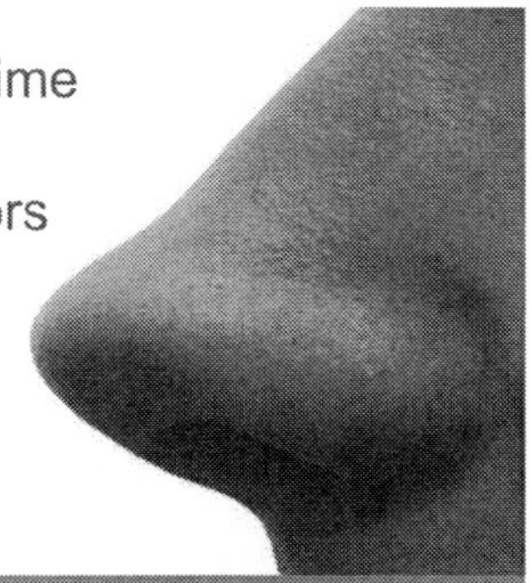

| ICAO UNITING AVIATION

Potential for impairment

- Potential acute symptoms
  - as a result of exposure to oil or hydraulic fluid fumes
- Impact on crew member performance
- Symptoms may develop slowly
  - degraded performance may not be initially obvious

ICAO UNITING AVIATION

Who should be trained?

ICAO UNITING AVIATION

What should training cover?

- SOPs to recognize and respond to fume events
- Reporting of fume events

ICAO UNITING AVIATION

What to include in cabin crew training

- Location/source, type and intensity of fumes
- Communication
- Passenger and cabin management
- Post-event procedures
- Applicable documentation

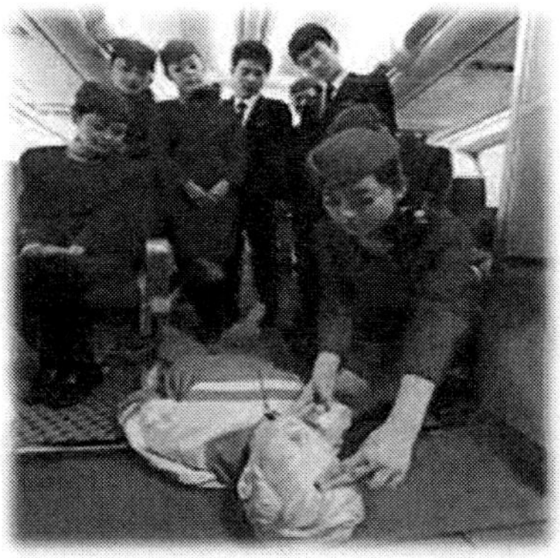

ICAO UNITING AVIATION

Standardized reporting: why?

- Reporting enables operator to document events
- Crew reporting is essential to assist AMT
  - flight and cabin crew must provide detailed and accurate information following event
- Standardized reporting assists in systematically gathering information
- Reporting at international level
  - ICAO developed standardized reporting form

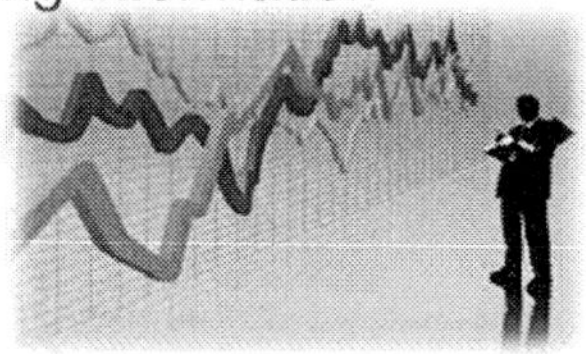

| ICAO UNITING AVIATION

Investigating events: what to look for?

- General information on event flight
  - data from reporting form
- Relevant documentation
  - maintenance logs, injury reporting forms
- Aircraft information
- Human performance
- Additional information
  - medical reports, first responder reports
- Interviews

| ICAO UNITING AVIATION

Conclusion

- Fume events should be addressed in training
  - Crew awareness
  - Response to event
- Reporting is key
  - Identifying root cause
  - Monitoring issue
- Investigations can lead to improvements
  - Training
  - SOPs

ICAO UNITING AVIATION

mmaurino@icao.int

www.icao.int/cabinsafety

ISD's Identifying & Tackling the problem

Noel Houlihan ***Safety Training Officer***

Aer Lingus

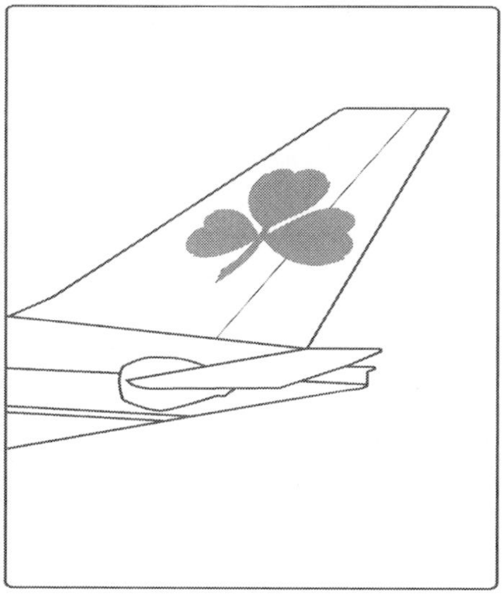

Who are we

- Aer Lingus is the National Carrier of Ireland
- Founded in 1936
- We are 80 years old this year

Aer Lingus

Who are we

- We are a medium sized carrier operating an all Airbus fleet
- We have 48 aircraft & employ 4000 + people
- We have 4 operating bases, throughout Ireland
- We operate both long & short haul routes carrying over 11 million passengers
- In August 2015 we became part of the IAG Group

Aer Lingus

Training Academy

- Training is centralised in our Dublin Training Academy
- We operate 8 classrooms, an A320 mock up, various door & slide trainers along with a fire training unit.

Aer Lingus

Incidents across the globe

Aer Lingus

The Aer Lingus way!

Aer Lingus

The Story Begins

- Back in the early 2000's Aer Lingus started looking at its door procedures for the Airbus, following an increase in slide deployments.
- Following a lengthy review it was highlighted that the door procedures had not evolved to meet the demands of a modern aircraft fleet.
- Robust procedures were developed that involved the entire crew, not just the cabin crew

Aer Lingus

Training

- The new procedures were reinforced by taking a radical approach to door training:
  - Door training would now be conducted annually during recurrent training, rather than being conducted on a triennial basis
  - This had a significant effect with crew confidence with doors growing significantly.
  - Practical assessment pass rates on doors grew to 98%
  - Crew confidence with evacuation commands also increased

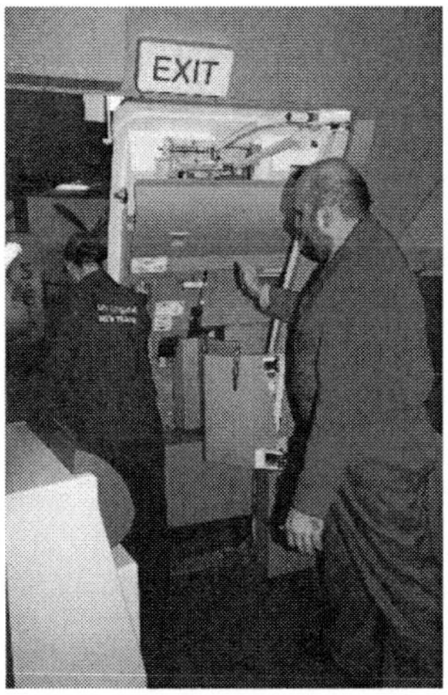

Aer Lingus

Statistics

- The results spoke for themselves, with deployments at an all time low!
- Events which did occur were caused by human error or as a result of a breach of procedure
- Until a sudden spike in 2015

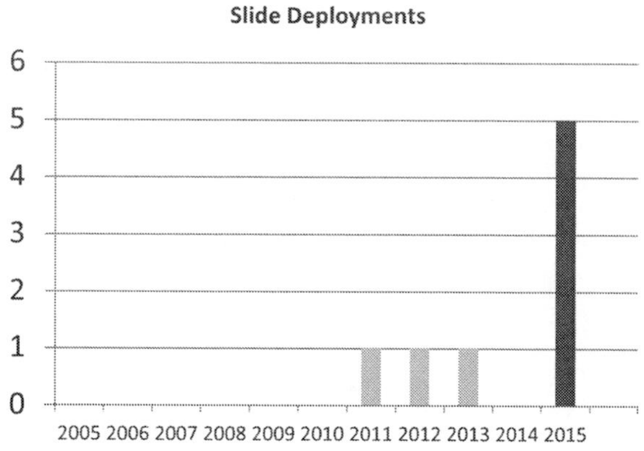

Aer Lingus

What went wrong

- With such a sharp increase in incidents we had to establish very quickly what had gone wrong.
- Following each incident the crews involved were debriefed thoroughly with the aim of finding out what went wrong
- The training team set about looking at the causes and feeding the information into our SMS system
- The safety management system became key to establishing the cause
- Working together the Corporate Risk Office, Flight Ops and Cabin Crew Training quickly began to shed light on the situation

Aer Lingus

The Findings

- Our procedures were robust, with little room for error
- The crew knew the procedures and were confident with them
- The crew could perfectly complete arming & disarming procedures when asked to do so during retraining
- To find the root cause we needed to dig deeper, so we started to look at the wider picture, after all what crew member comes to work with the intention of deploying a slide.

Aer Lingus

The Findings

- 2015 was one of our largest growth years in history with large quantities of new crew joining, was this to blame?
- Passenger numbers had soared and with it increased disruption at our main hub, shorter turn arounds, and increased pressure on crews.
- Crews involved were re-interviewed and the incidents were reassessed
- Very quickly we established that the key factor was distraction
  - None of the incidents occurred during normal operations, it was always during disruption or at a time when someone outside the crew did something unexpected
  - Only one incident involved a new crew member, but again distraction was the key factor.

Aer Lingus

CDG incident 2015

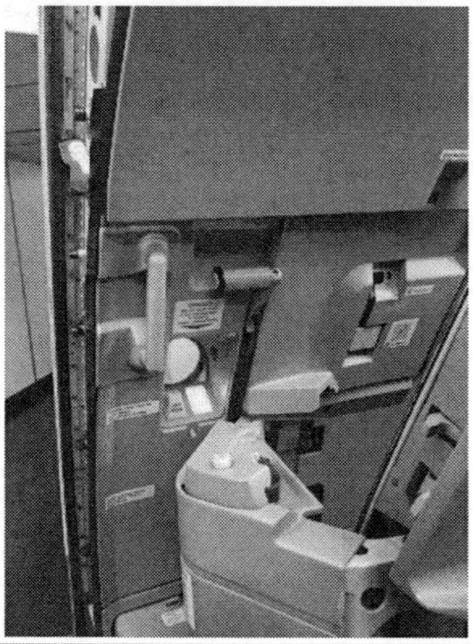

Aer Lingus

The Solution

- It was now clear that the only solution could be a training one, so we set about developing an awareness program and new training program for 2016
- Awareness was raised by sharing the incidents with the crew body and by including discussions on doors & distraction in pre flight briefings

Aer Lingus

The Solution

- In practical training we have added in causes of distraction to arming and disaming procedures
- We have added a Continuous Professional Development module on slide deployments to recurrent training for all crew
- We have developed a new process in training for crew called Slow Down

Aer Lingus

The solution

- We have got the crew involved to talk about the incidents with their colleagues, rather than being embarrised as crew often are
- These talks have been in class and have opened everyone's eyes to personal element of an incident and how the person feels
- We have encouraged enhanced hazard reporting so that further information can be fed through SMS
- We have developed similar modules for new intake crew to ensure clear understanding and to prevent cross contamination from other carriers
- We continue to review and analyse the data as it arrives so that further improvements can be made into the future!

Aer Lingus

Aer Lingus

Evacuate!

Evacuate!

Evacuate!

Page 1

Kris Hutchings
Manager, Inflight Safety & Quality

WESTJET

WestJet

WestJet Snapshot

- Based in Calgary, Alberta, Canada
- Over 160 million guests flown in 20 years
- WestJet operates a modern fleet comprised of Boeing Next-Generation 737-600, 737-700 & 737-800 and Boeing 767-300ER Aircraft.
- WestJet Encore operates Bombardier Q400 NextGen turboprops.

Safety Culture

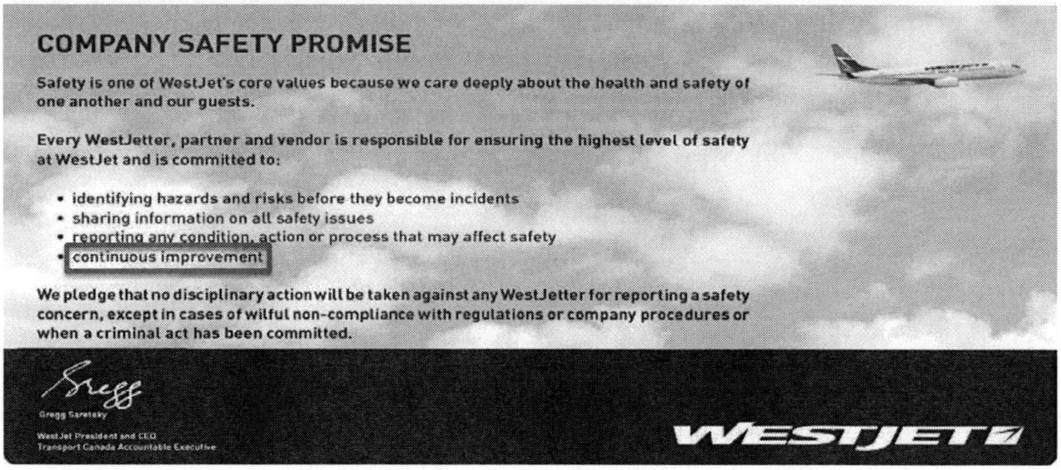

Page 3

WESTJET

Why change your evacuation procedures?

Page 4

WESTJET

Introduction of wide-body aircraft

- Welcome B767-300ER to the WestJet family
- WestJet's first experience with having flight attendants being qualified on two aircraft types
- Continuity of commands and evacuation flow

Transport Canada – New regulations

725.124(58) – Training for Flight Attendants Assigned To Open More Than One Exit During An Aeroplane Emergency Evacuation

(*a*) Initial and annual training for flight attendants assigned to open more than one exit during an aeroplane emergency evacuation shall include:

(i) methods of passenger flow control management, including that the flight attendant is to:

(A) evaluate passenger use of exits and direct passengers to another exit to increase the number of passengers evacuating the aeroplane;

(B) continually assess the condition of exits and direct passengers to available exits; and

(C) demonstrate passenger flow control management techniques during evacuation drills including the signals and commands necessary to maximize passenger evacuation from the aeroplane.

Lessons learned from an elevated incident

Page 7

"This is your Captain speaking..."

"...we will be diverting to Winnipeg and expect to be landing in approximately 20 minutes"

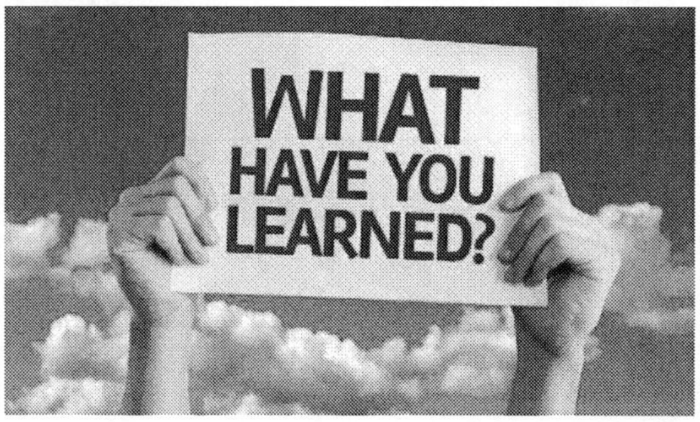

Page 8

Lessons learned

- Time required to open both aft exits with a single flight attendant
- Content of evacuation commands
- Flow of passengers during an evacuation
- Directions given to ABPs outside the aircraft

WESTJET

Impact to Flight Attendants

WESTJET

Revised Evacuation Flow

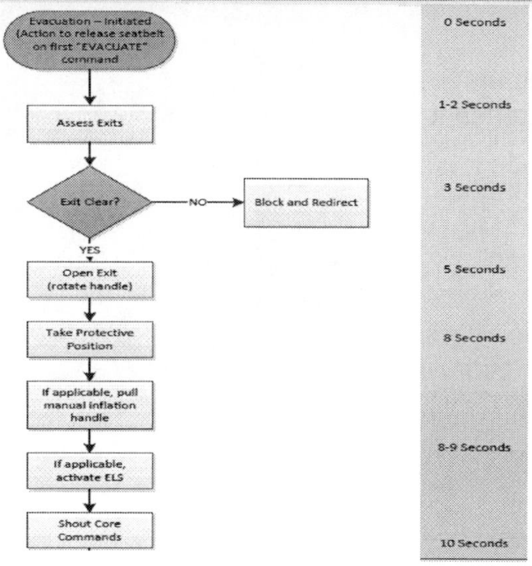

Key element

Initiate door opening within 5 seconds

Page 11

WESTJET

Old Protective Positions

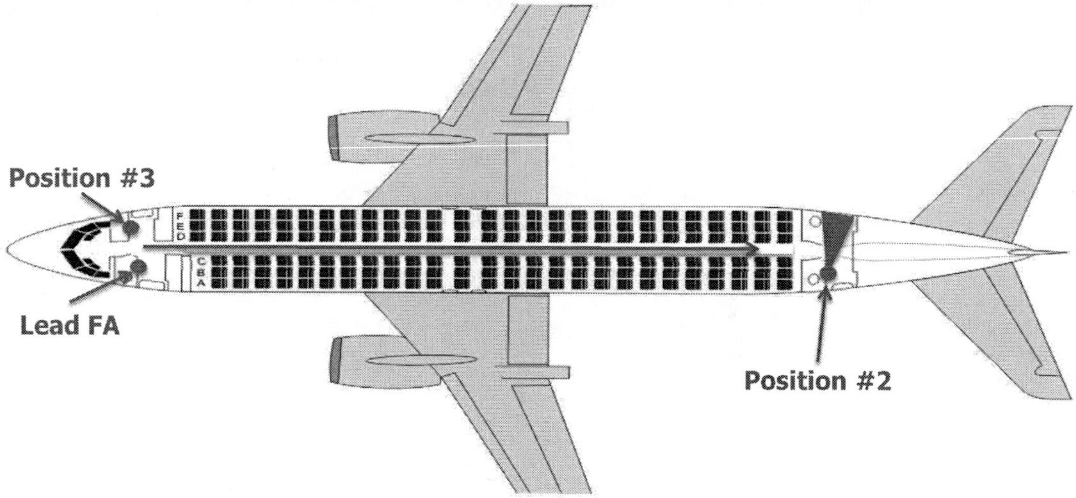

Page 12

WESTJET

New Protective Positions

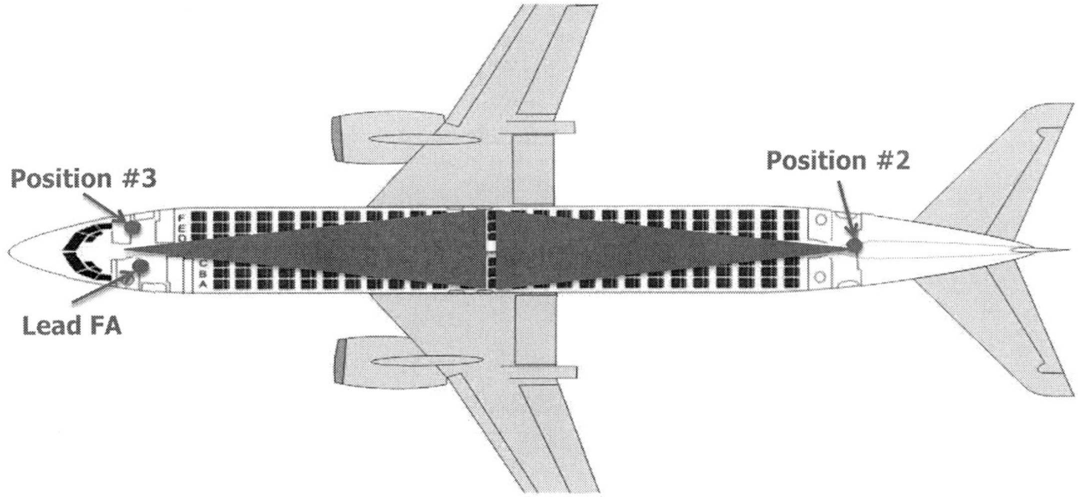

Page 13

WESTJET

Revised Evacuation Commands

Leave Everything!

Get Out!

Page 14

WESTJET

Training Resources

- Communication strategy
- Coordinated manual revision effective date
- Drill workshops
- Videos of new evacuation procedures

Page 15

Summary

- Continuous improvement
- Change isn't always scary
- Start with "why"

Questions?

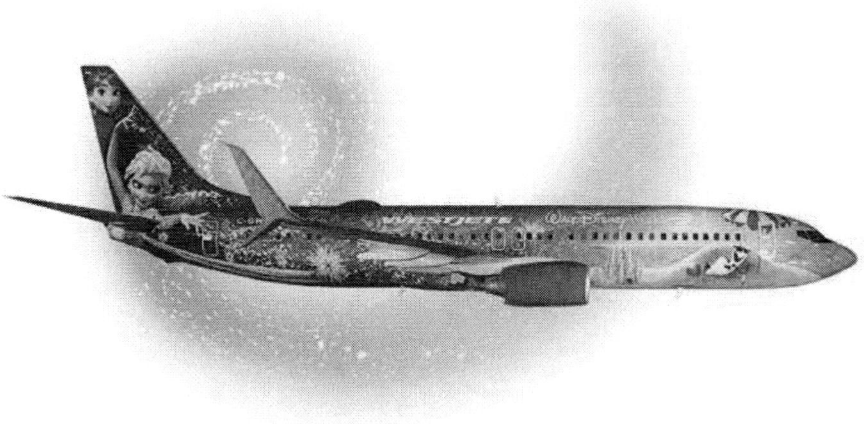

Page 17

Thank you!

Kris Hutchings

Kris.hutchings@westjet.com

Page 18

RISK MANAGEMENT: IT'S ALL IN YOUR HEAD

Dr Shari Frisinger, President, CornerStone Strategies LLC

Colette Hillary, Aviation Training Consultant

PERSONAL NEEDS

Fatigue

- ❖ Rest Related Fatigue – Acute, Chronic, Cumulative, Transient

Environmental

Emotional

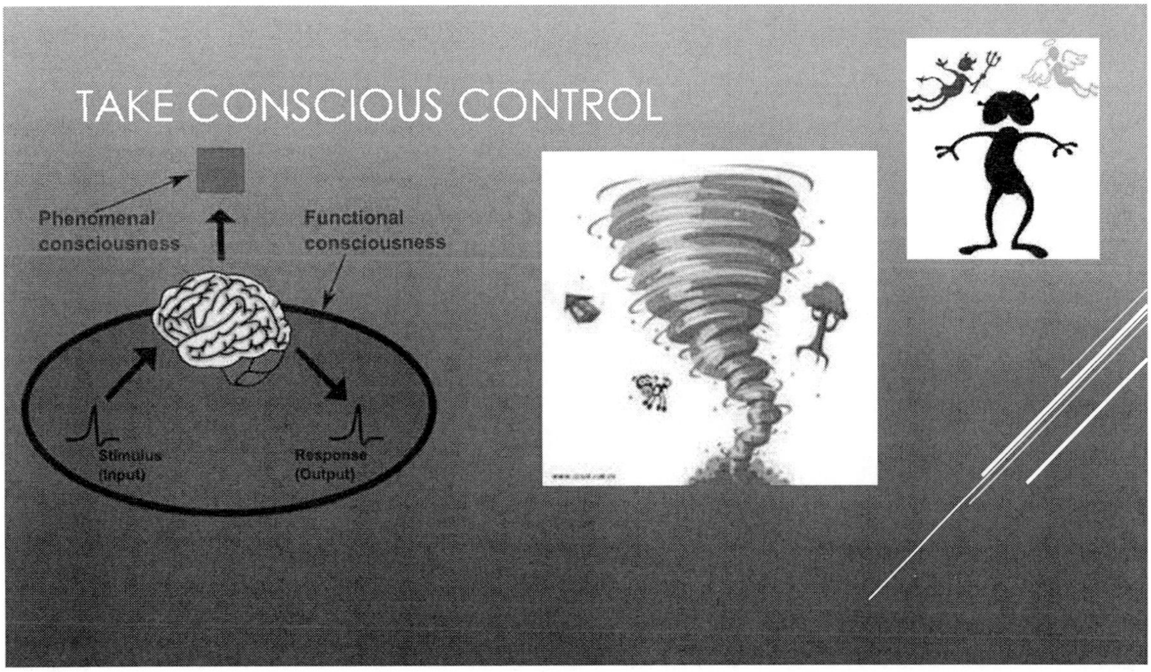
TAKE CONSCIOUS CONTROL
Phenomenal consciousness
Functional consciousness
Stimulus (Input)
Response (Output)

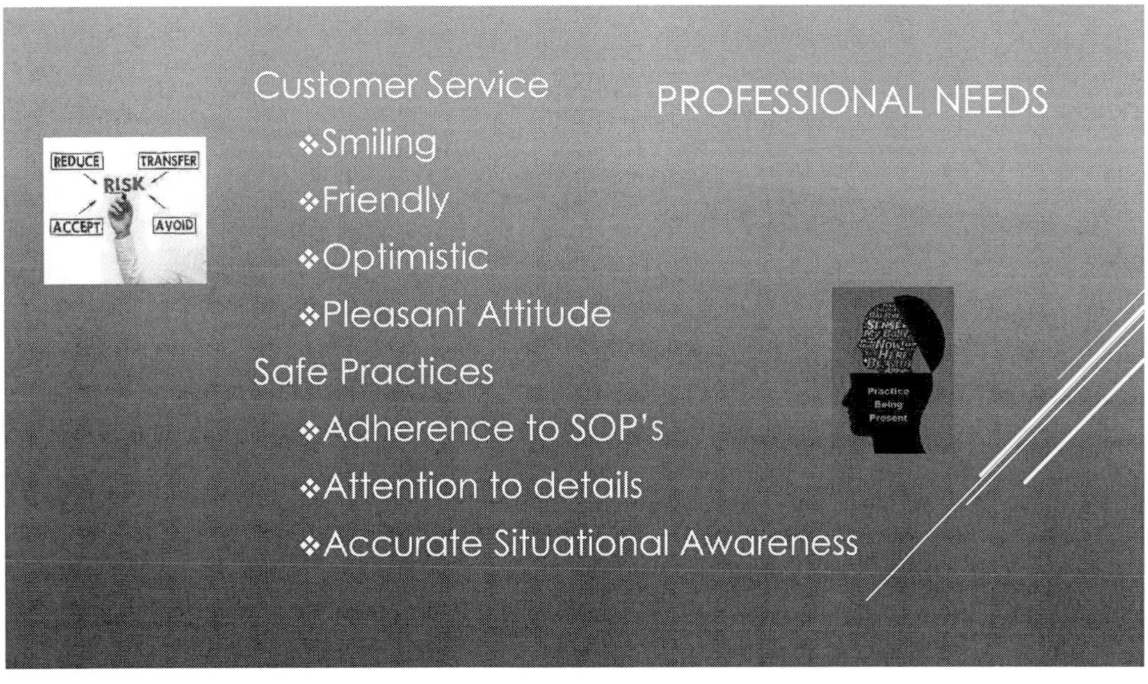
Customer Service
❖Smiling
❖Friendly
❖Optimistic
❖Pleasant Attitude
Safe Practices
❖Adherence to SOP's
❖Attention to details
❖Accurate Situational Awareness
PROFESSIONAL NEEDS
REDUCE
TRANSFER
RISK
ACCEPT
AVOID
Practice Being Present

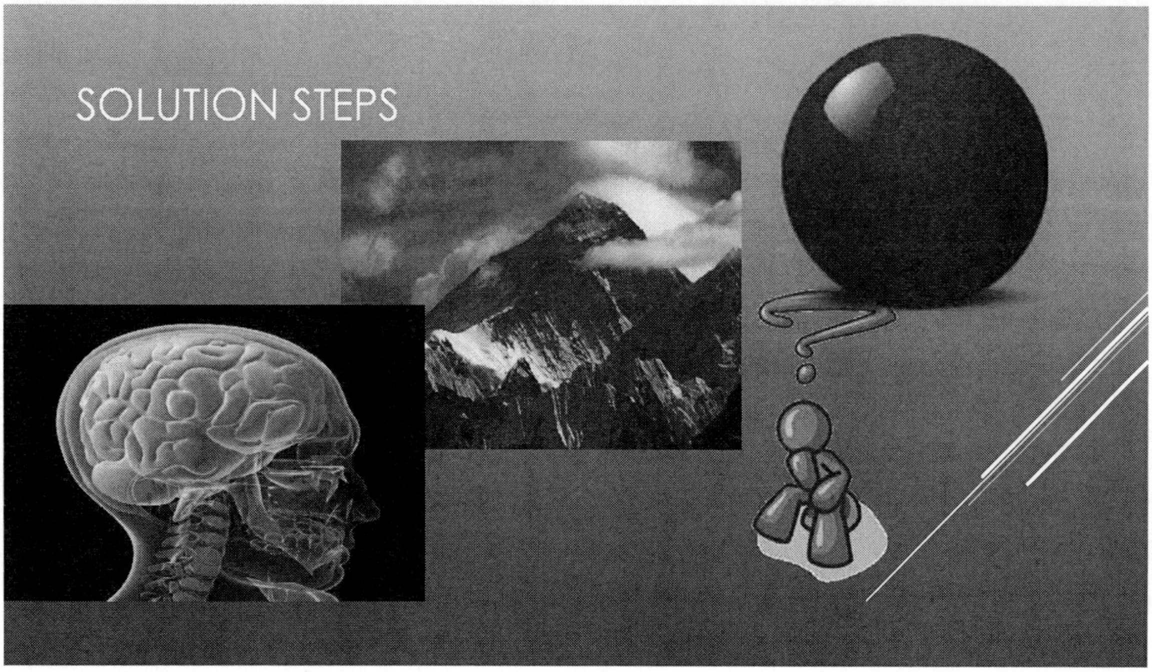
SOLUTION STEPS

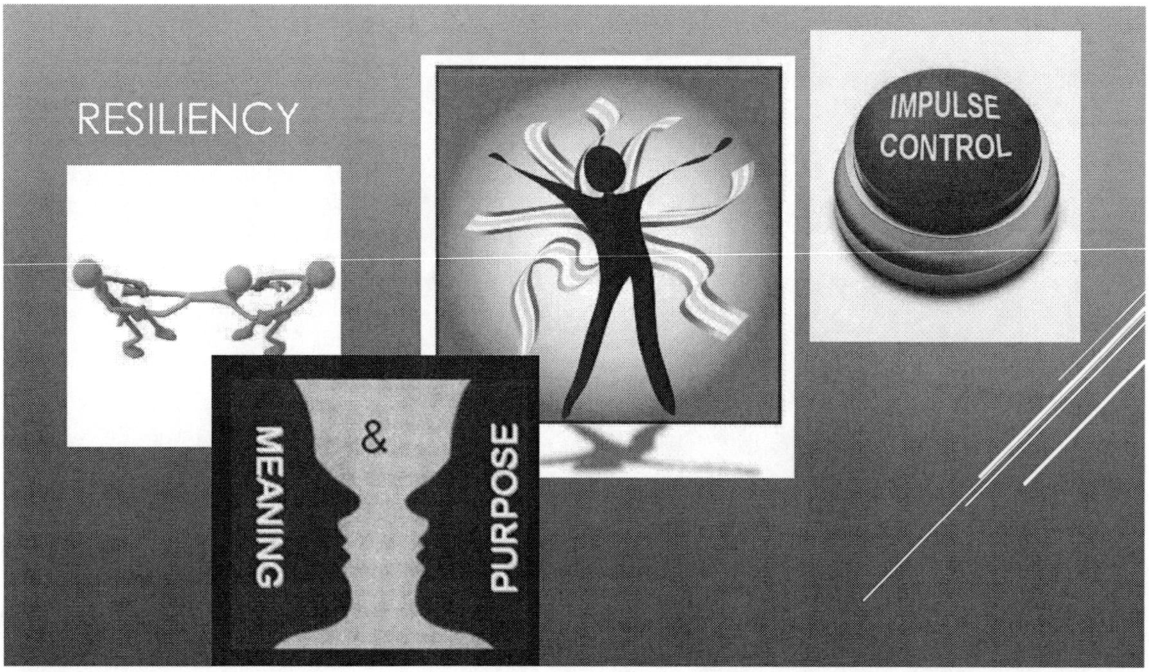
RESILIENCY
IMPULSE CONTROL
MEANING
&
PURPOSE

SELF-REGULATION

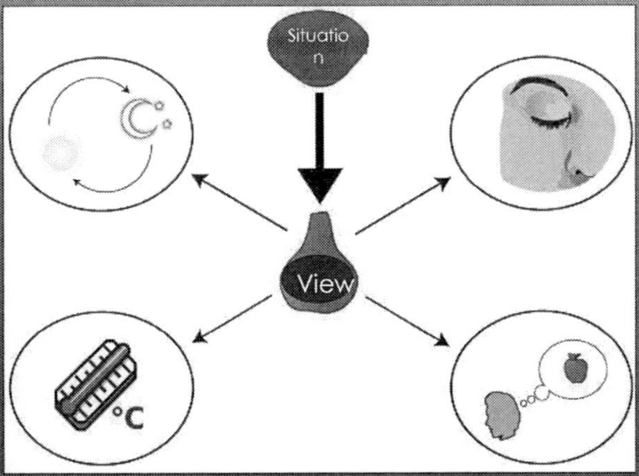

THE CONSEQUENCES

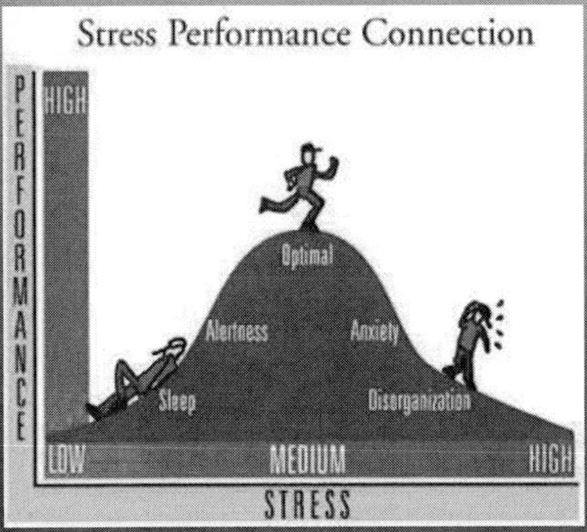

- Disguises/hides mental clarity, degrades performance …

IF YOU WOULD LIKE THIS INFORMATION

In a checklist format you can share with your crew and use in your training, contact us:

Dr. Shari Frisinger
281.992.4136
Shari@CStoneStratLLC.com
Connect through Linked-In
Or Twitter (@shari_f)

Colette Hilliary
404.989.4425
Colette.Hilliary@gmail.com

WELCOME
TO
MAGERVILLE
Guarantee Training Results
Using Mager's CRI
Mager Consortium
Common Sense in Workplace Performance

Dr. Robert Mager

Three predictable results of using CRI

- Training takes less time
- Results are **GUARANTEED**!
- Learning curve is **ELIMINATED!**

Mager Consortium
Common Sense in Workplace Performance

CRI Eliminates the Learning Curve

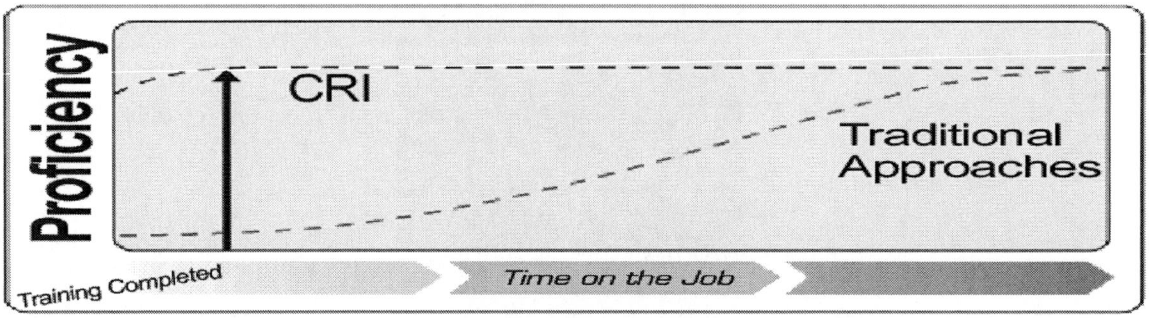

Mager Consortium
Common Sense in Workplace Performance

Before and After

Existing Training	CRI Training
5 weeks	5 days
50% of trainer's time	5% of trainer's time
92% accuracy	98% accuracy

Mager Consortium
Common Sense in Workplace Performance

It's All About *Self-Efficacy*

Each learner must ***demonstrate*** that he/she can meet pre-specified criteria for ***each skill*** the course is designed to teach

Mager Consortium
Common Sense in Workplace Performance

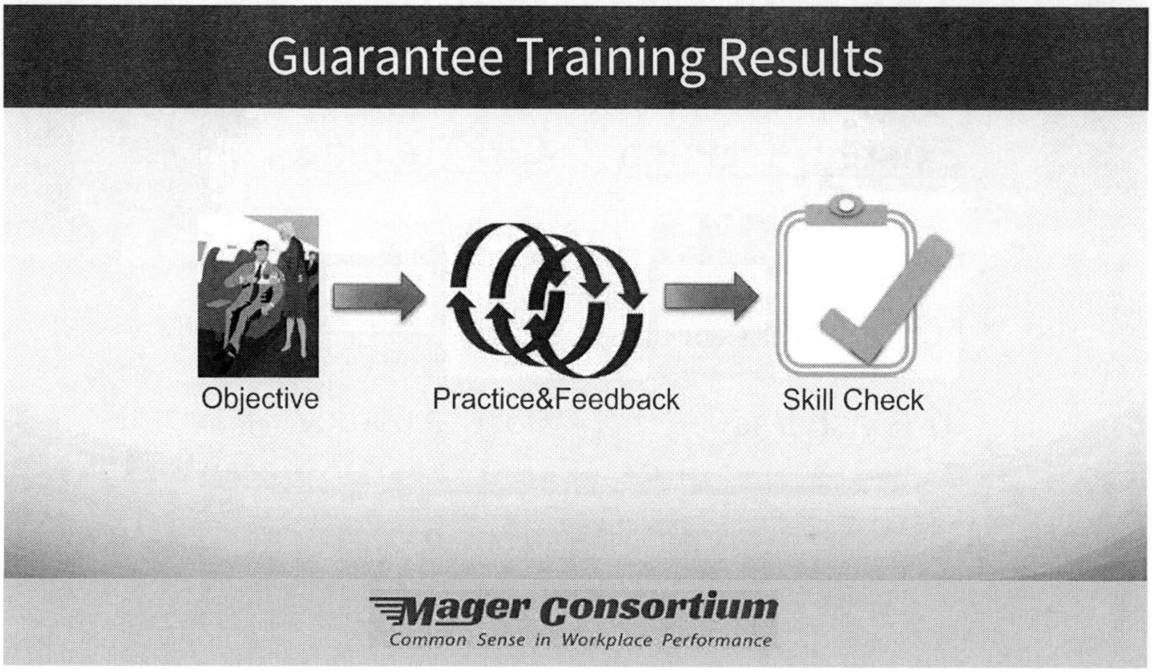
Guarantee Training Results
Objective
Practice&Feedback
Skill Check
Mager Consortium
Common Sense in Workplace Performance

Training Components
Performance Objective
?
What's In It For Me (WIIFM)
Practice and Feedback
Skill Check
Mager Consortium
Common Sense in Workplace Performance

Focus

Skill

not

Information

Emphasis

Performance

not

Content

Evaluation

Skill Checks

not

Knowledge Tests

Requirements for Performance:

Skill Self-Efficacy	→	Training Responsibility
Opportunity to Perform Supportive Environment	→	Management Responsibility

Mager Consortium

Common Sense in Workplace Performance

Performance Problem Causes

1. Lack of ***skill***
2. Lack of motivation ***(will)***
3. Other obstacles ***(hills)***

Mager Consortium

Common Sense in Workplace Performance

CRI Culture

The goal is

Simplicity

&

Clarity

Mager Consortium
Common Sense in Workplace Performance

CRI: The Flexible Foundation for

- **ALL forms of instruction**

  self-paced, group-paced or blended

- **Any delivery method or with any combination of media**

  print, video, audio, CBT, e-learning via the internet, intranet, cloud, etc.

Mager Consortium
Common Sense in Workplace Performance

Key Business Systems

- Recruiting/hiring
- Training
- Measuring & evaluating performance
- Career progression & succession planning

Mager Consortium
Common Sense in Workplace Performance

Criterion-Referenced Instruction

Your results should be

SPECTACULAR!!

Mager Consortium
Common Sense in Workplace Performance

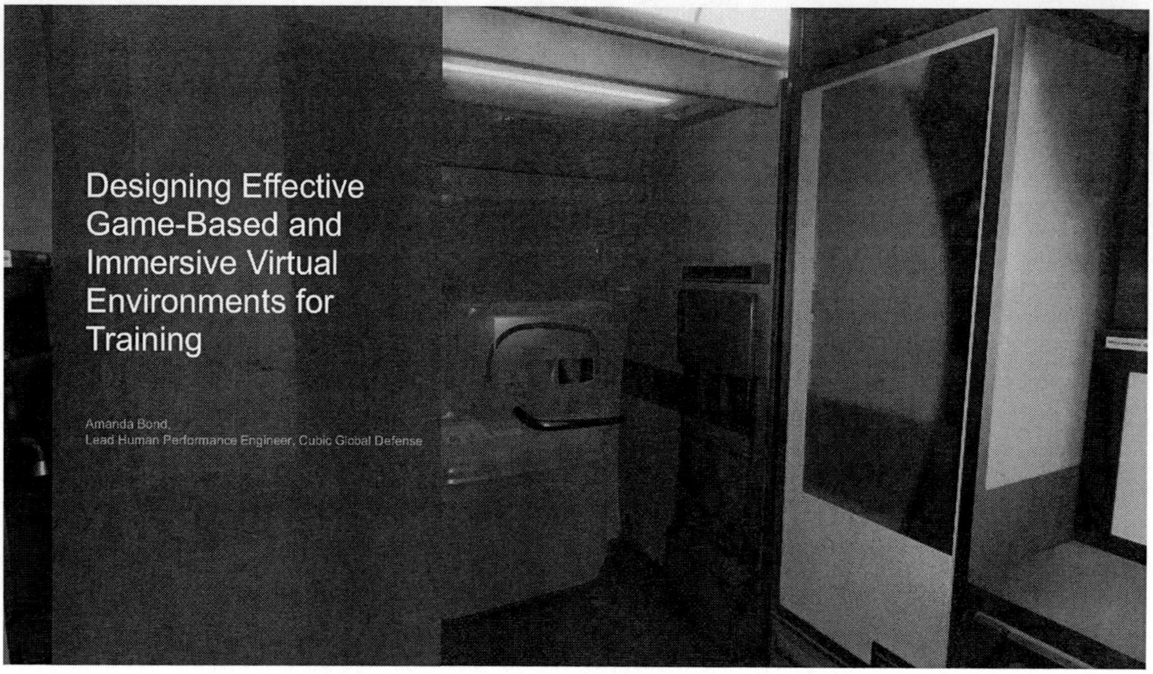

Agenda

2

- Introductions
- Games, Simulation, and Training: Definitions and expectations
- How do I know if my organization could benefit from using immersive/game-based training?
- Designing instruction within an immersive virtual environment
- Designing the immersive virtual environment
- Gamification strategies

CUBIC

Cubic Proprietary

3

Introductions

Cubic Proprietary

4

About Me

Lead Human Performance Engineer, Cubic Simulation Systems, Inc.

15 years as a Human Factors Psychologist/Engineer

Experience across commercial and defense applications in Training, Evaluation, & Design

Microsoft Research Female Academic All-Star: Serious Games

CUBIC

Cubic Proprietary

About Cubic Corporation

Company Profile

Cubic was founded in 1951 in San Diego, CA, USA

- **54 years as a publicly traded company**
  - NYSE: CUB
- **Global Operations**
  - 8,000 employees at 130 locations worldwide
- **Primary Markets**
- **Defense**
  - Air Combat Training
  - Ground Combat Training
  - Virtual and Gamed Based Training
  - Special Operations & Intelligence
  - Communication Products
  - Cyber Training
  - LVC Training and Performance Assessment
  - Mission Support Services
- **Transportation Systems**
  - NextCity
  - Automatic Fare Collection
  - Operations Support
  - Maintenance Services

CUBIC

Cubic Simulation Systems, Inc. (CSSI)

Summary

Established in 2012, **Cubic Simulation Systems, Inc.(CSSI)** is the **expert** in the industry of effective learning products, solutions and human performance.

CSSI is a **cost-competitive** leading provider of **innovative** and **effective training**, **modeling & simulation**, and other **performance improvement solutions**.

CSSI creates challenge-based, **collaborative** learning solutions that provoke critical thinking, distinctive communication and problem solving through the creation of **game based applications** containing solo and group content required to **achieve complex goals** tied to a storyline.

Mission

We create training innovation through advanced products, services and solutions by positioning the latest instructional strategies and technologies for individuals, businesses and governments leading to measurable performance improvement.

Our People

The over 100 employees of CSSI have the following remarkable credentials:

- Over 25% have Advanced Degrees
- **35%** have an average of **21** years service in the Military
- Over **45%** of the Game Development staff have participated in creating multiple major game titles and collectively shipped **25** titles
- 10% of the entire staff speak more than one language fluently

Mobile Apps * Game-Based Learning * Learning Analytics * Gesture-Based Computing

CUBIC

7

Games, Simulation, and Training: Definitions and Expectations

Cubic Proprietary

8

Definitions

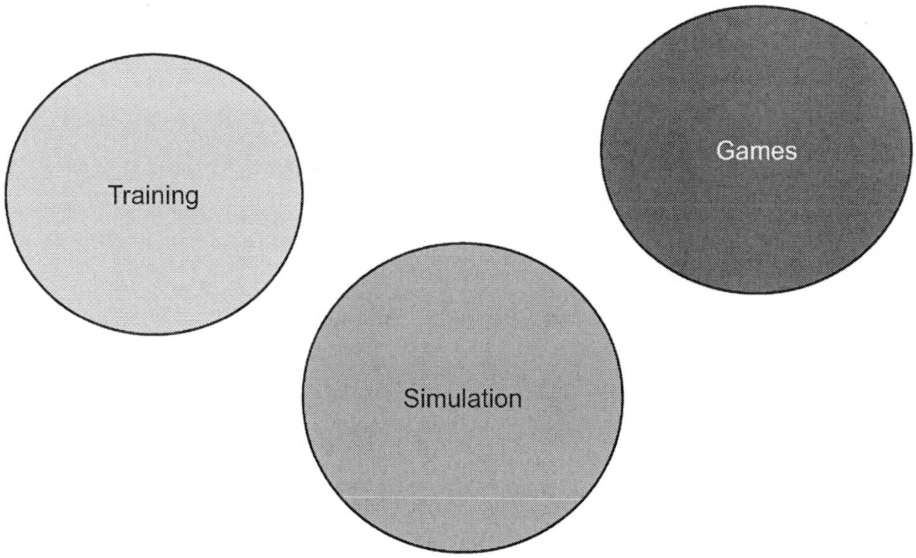

CUBIC

Cubic Proprietary

9

Definitions

A **game** is:

- interactive
- Inherently motivating
- rule-governed
- Where an individual progresses to achieve a goal
- The primary purpose is fun
- Games can be live or computer
- Games can be anything from playing cards to laser tag, either real or virtualized
- Skill, strategy, chance, or a combination
- Single-player or multi-player

CUBIC

Cubic Proprietary

10

Definitions

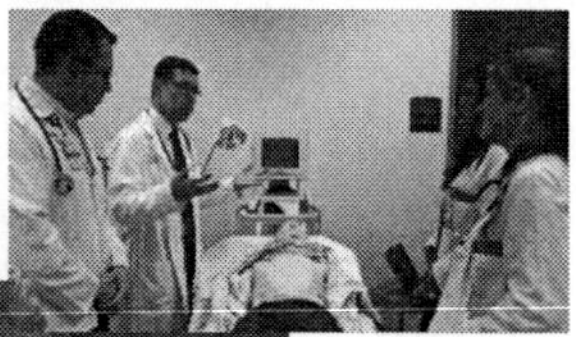

Training is:

- Organized action of teaching (or learning) a particular skill or behavior
- Goal of improving one's knowledge or performance
- Agnostic of delivery method
- Purposefully assessed (though may not be graded!)
- Includes feedback and opportunity for remediation

Note: Practice doesn't make perfect: but deliberate practice does!

Deliberate practice is focused on performing the task well, and improving each time through feedback

"Tuba in a closet"

CUBIC

Cubic Proprietary

Definitions

11

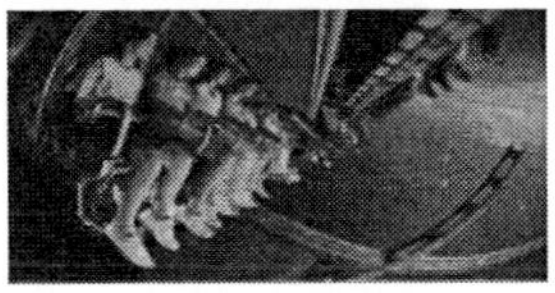

A **simulation** is:

- the imitation of the operation of a real-world system, process, or environment over time
- Often includes a model of the system's behavior, which is a body of information about the system

A model of behavior allows flexibility in simulation – dynamic simulation rather than linear processes

Cubic Proprietary

Definitions

12

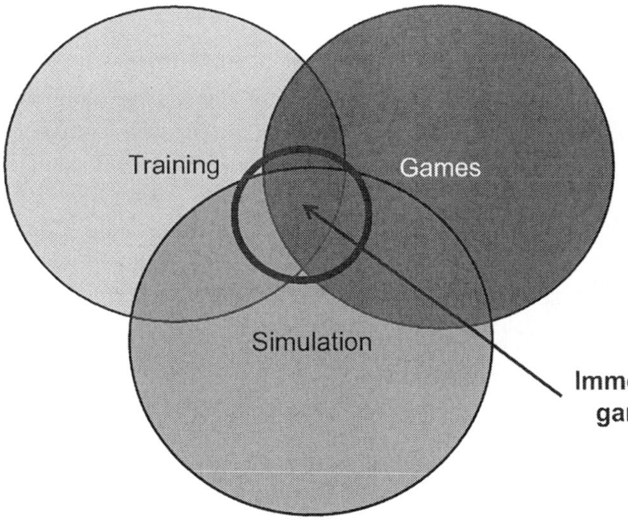

Cubic Proprietary

Definitions

13

Educational Games	Virtual Reality	Social Impact Games
Learning Games	Virtual Worlds	Persuasive Games
Training Games	Alternative Purpose Games	Games for Change
Games for Health	Digital Game-Based Learning	Synthetic Learning Environments
Health Games	Simulations	Game-based [domain]
Simulation/Simulators	Immersive Learning	Games for Good

A variety of labels are used to describe game-based immersive training, both correct and incorrect.

CUBIC

Cubic Proprietary

Definitions

14

The "FlatWorld Wide Area Mixed Reality" demonstration by the United States Army

So what does "**Immersive**" mean?

- A sense of "being there," even when you aren't
- Promotes better, faster learning
- Attention is focused on the immersed space
  - Encoding is happening for the immersed space
- Feeling of "immersiveness" ("presence") comes from a variety of factors (Witmer & Singer, 1994):
  - Control
  - Sensory (Visual/auditory/haptic) cues
  - Feedback

CUBIC

Cubic Proprietary

15

Definitions

Examples of immersive game-based training

Immersive environments can consist of:

- Virtual art model
- Audio tones/cues/etc.
- Non-player characters
- Simulation model of component behavior
- Varying degrees of realism & complexity
- Haptic devices
- Physical artifacts such as yokes
- Vibratory feedback in input devices such as mice
- Combinations of the above

CUBIC

Cubic Proprietary

16

How do I know if I could Benefit from Immersive/Game-Based Training?

Cubic Proprietary

17

Benefits of Game-Based Training

Make the time spent in physical simulators more *effective*.

When trainees learn and practice first in the game, the first time in the simulator, s/he …

- … will spend less time struggling with what to do because they are **already familiar** with the physical environment.
- … will receive instruction **tailored** to their needs in the simulator.

This approach is effective because of **schematic processing**. Since s/he has already developed a foundation for the operational environment and the tasks, they will learn more quickly.

This can decrease the time needed in physical simulators

CUBIC

Cubic Proprietary

18

Benefits of Game-Based Training

Expose trainees to scenarios that are difficult— or even impossible — to train.

When trainees experience dangerous situations in the game, s/he…

- … can **practice skills and problem-solving techniques** that they would otherwise not be able to do…
  - Fire fighting
  - Complex medical emergencies
  - Unruly, violent passengers
- … can drill emergency scenarios over and over until they are **trained to automaticity**.

The highly realistic, immersive environment encourages **transfer of training**.

CUBIC

Cubic Proprietary

19

Benefits of Game-Based Training

Game-based training can be adaptive based on individual learner performance

Training centers can then…

- Recreate incident conditions for practice of regulations/guidelines
- Make new scenarios to apply learner knowledge in ambiguous situations
- Create complex, reactive, multi-layered scenarios and/or dynamic scenarios – e.g., CRM training

Practice Situational Awareness, teamwork, collaboration, decision-making, communication, etc.

Operations
Maintenance
Passengers
Time Constraints
Culture
Automation
Traffic
Weather
Workload
Equipment
ATC
Crew
Fatigue

CUBIC

Cubic Proprietary

20

Benefits of Game-Based Training

Obtain in-depth data for analytics

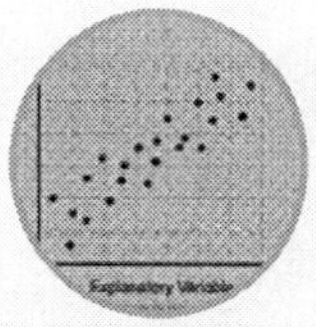

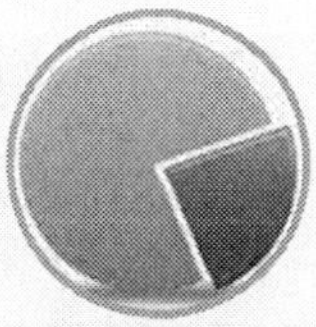

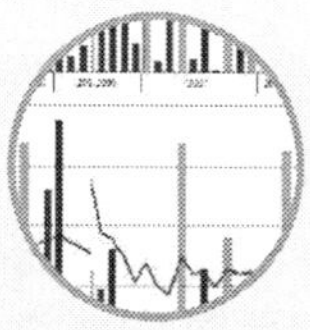

The trainee's performance can be tracked and analyzed *based on performance as well as knowledge…*

… to determine gaps in knowledge and skill so that they may be addressed (either by a virtual or real instructor).

… so the instructors will know what they need to spend time on with trainees, making the most of their time.

… to determine what **training they need** and what they can skip to optimize their time at the training center.

CUBIC

Cubic Proprietary

Schemas in Processing

21

If it sounds like a duck, but doesn't look like a duck ...

"A schema is a script and/or structure for common occurrences that run on auto-pilot."

We use schemas for:

- Tackling new problems
- Sensemaking
- Learning

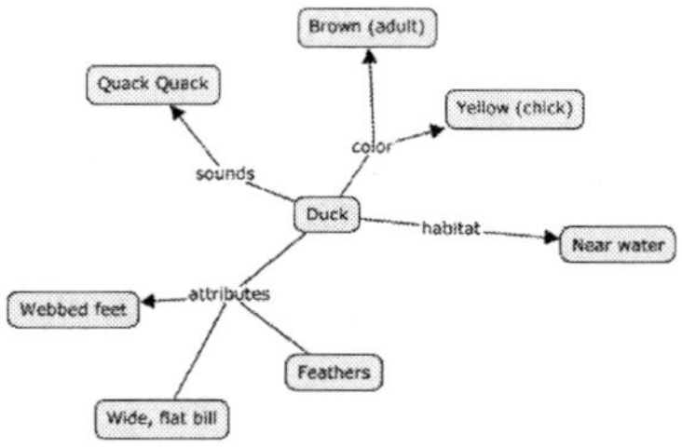

Cubic Proprietary

Experiential Learning

22

Caveats:

- The learner must be willing to be actively involved in the experience
- The learner must be able to reflect on the experience
- The learner must possess and use analytical skills to conceptualize the experience
- The learner must possess decision making and problem solving skills in order to use the new ideas gained from the experience

Kolb, 1984

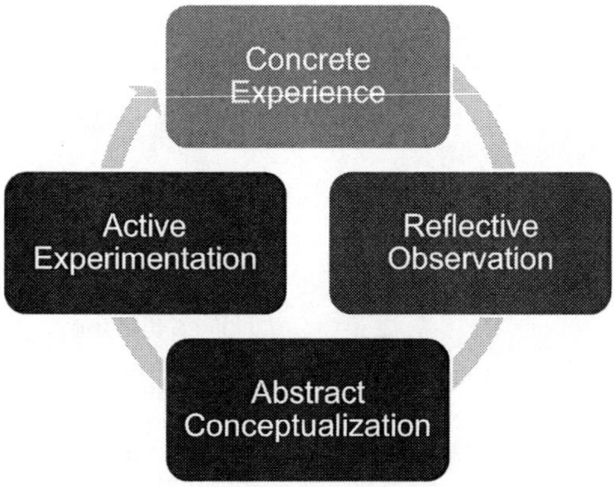

Cubic Proprietary

Building Expertise

23

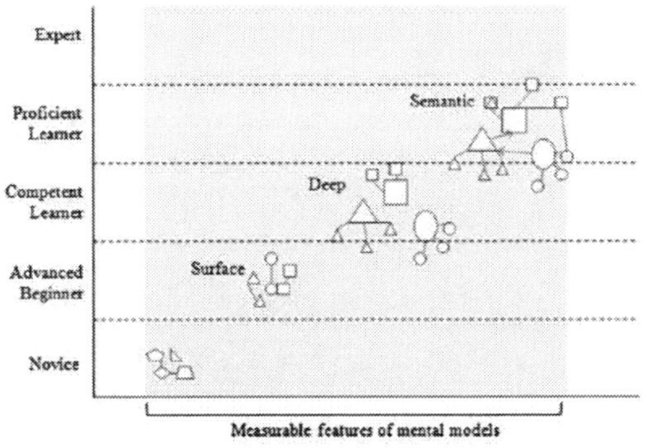

Levels of expertise expressed as mental model maps

- Expert-Novice differences
- Deliberate practice
- Schema-building
- Deep knowledge

CUBIC

Cubic Proprietary

Building Expertise

24

- What is an expert?
  - 10,000 hours (Ericsson, 1990)
  - Office worker: 40 hours week x 52 weeks per year = 2,080 hours/year
    - 5 years of doing variants of the same task to become an expert
    - ***Note:*** *of deliberate practice*
  - U.S.: 80 flight hours a month typical for cabin crew
    - 80 hours x 12 months = 960
    - 10+ years to become an expert
  - How long is the average airline contract?

CUBIC

Cubic Proprietary

Return on Investment

25

Increase Training Efficacy

- Stizman (2011) Meta Analysis found
  - ✓ 20% increase in Self Efficacy (confidence)
  - ✓ 11% increase in Declarative knowledge
  - ✓ 14% increase in Procedural knowledge
  - ✓ 9% increase in Retention
- Other studies show even larger gains in specific domains
- E.g., Wouters, van Nimwegen, van Oostendorp, & van der Spek, 2013

CUBIC

Cubic Proprietary

Return on Investment

26

Better training **efficacy** means better training **efficiency**

- Operationally:
- Opportunity for more footpath on expensive simulators
  - Reduce iterations on the simulators to put more trainees through
- Less wear & tear on existing simulators
- Standardized training
- Instructors can focus their valuable time on just the trainee's specific needs
- Get trainees to the line faster
- Trainees can review at their leisure; less likely to fail

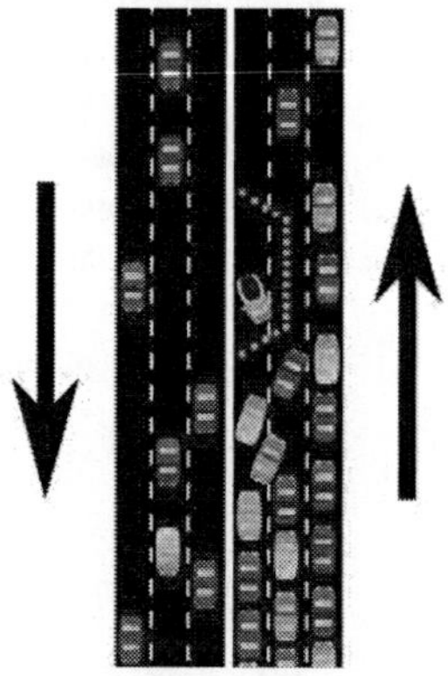

CUBIC

Cubic Proprietary

Return on Investment

27

Consider the Total Human Cost of Ownership

Consider non-operational and non-course-development costs of the hiring/training

- Hiring cabin crew
- Attrition of cabin crew
- Incidents
- Customer Service impacts
- Travel...
- Lodging....

CUBIC

Cubic Proprietary

28

Designing instruction within an immersive virtual environment

Cubic Proprietary

29

Designing Effective Immersive GBT

First and foremost: What do you **need** to train?

Distinctly different than *want* to train

How do you know?

CUBIC

Cubic Proprietary

30

Analyze

- There are multiple analyses that can be conducted in order to design the best training for the specific domain/task/purpose
  - Not all analyses are always needed – each uncovers a different type of information

- Organizational Analysis
- Work Domain Analysis
- Job/Duty/Task Analysis
- Sensory Task Analysis (For VEs)
- Cognitive Task Analysis (Complex Decisions)
- Current/Desired Performance Analysis
- Training/Current State Analysis
- Gap Analysis
- Media/Technology Analysis

CUBIC

Cubic Proprietary

31

Training Needs Artifacts

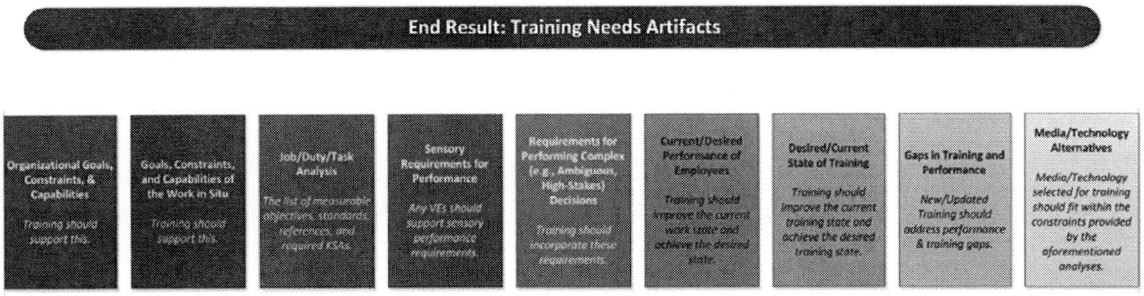

CUBIC

Cubic Proprietary

32

Task Analysis

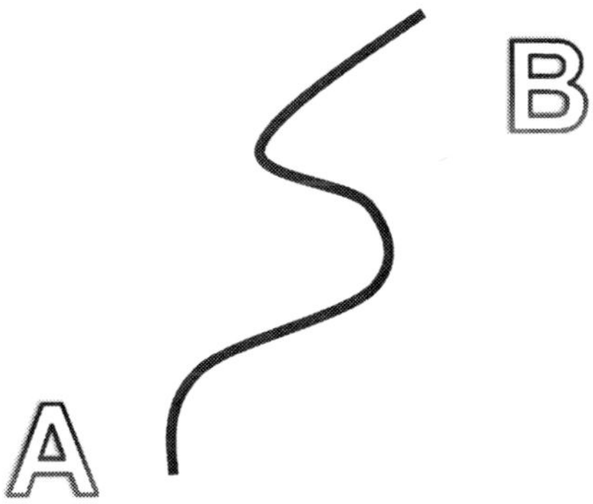

What is the correct path from A to B?

Task Analysis:
How a user gets from Point A to Point B

- The procedural step-by-step of a task

- Direct Observation Method
- Protocol Analysis/Talk Aloud Method
- … and many more

CUBIC

Cubic Proprietary

Task Analysis

33

Task	Action	Component	End State	Reactive	End State
PA: "Cabin Crew prepare all doors and cross check"	Audio				
Raise the Plastic Guard	Lift	L2 Plastic Guard	Up		
Arm the L2 door	Move	L2 Slide Arming Lever	Armed	L2 Left Girt Bar Indicator	Yellow
				L2 Right Girt Bar Indicator	Yellow
Lower the Plastic Guard	Lower	L2 Plastic guard	Down		
Verify Girt Bar Indicators	Verify	L2 Left Girt Bar Indicator	Yellow		
	Verify	L2 Right Girt Bar Indicator	Yellow		
Cross Check	Navigate		R2		
	Verify	R2 Slide Arming Lever	Armed		

CUBIC

Cubic Proprietary

Cognitive Task Analysis

34

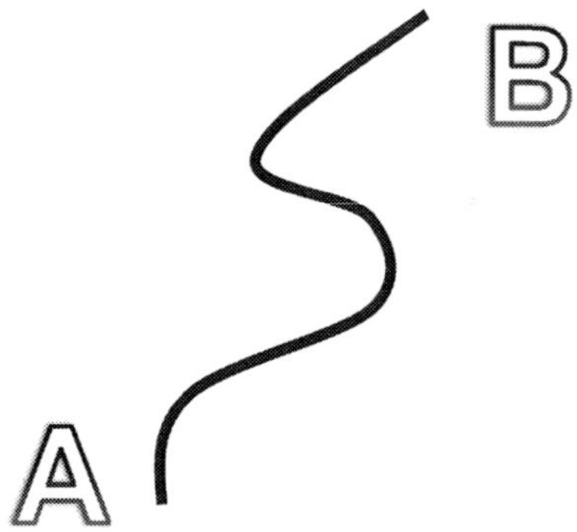

Cognitive Task Analysis: How a user *decides how* to get from Point A to Point B

- The cognitive step-by-step decisions and strategies of a task
- Critical Decision Method
- Applied Cognitive Task Analysis Method
- … and many more

How would you decide how to go from Point A to Point B?

CUBIC

Cubic Proprietary

Cognitive Task Analysis

35

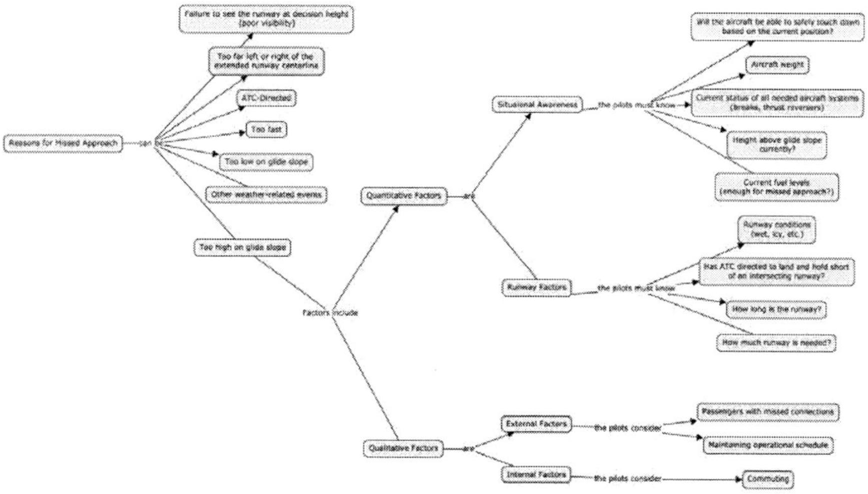

CUBIC

Cubic Proprietary

Sensory Task Analysis

36

Sensory Task Analysis: How a user detects their environment as they are getting from Point A to Point B

- Identifies sensory and perceptual cues in each stage of task completion
  - voluntary and involuntary
- E.g., visual cues in the environment for course of action decisions
- Can be employed embedded with other *cognitive task analysis* methods

CUBIC

Cubic Proprietary

Sensory Task Analysis

37

Task	Cue
Emergency evacuation – slide raft fully inflated	See Slide Raft fully inflate
Emergency evacuation – slide raft fully inflated	Hear Slide Raft fully inflate
Manual door opening – door fully open	Hear Gust Lock engaging
Emergency ditching – door status	See water level below door sill
Decompression	See cabin lights turn on
Lavatory Fire – feel if door is hot or not	Feel heat

Which cues are critical to the task and decision-making?
Which cues must be exact in the environment?
Which cues can be approximated (e.g., feeling heat)?

CUBIC

Cubic Proprietary

Assessment

38

- What do you need to ***assess***?
- Procedures? Memorization? Decision-Making? Techniques?
  - How a user gets from point A to Point B ***and***
  - How a user knows they're *successfully* getting from Point A to Point B
  - Measures of Performance
  - Measures of Efficacy
  - Etc.

CUBIC

Cubic Proprietary

Assessment

39

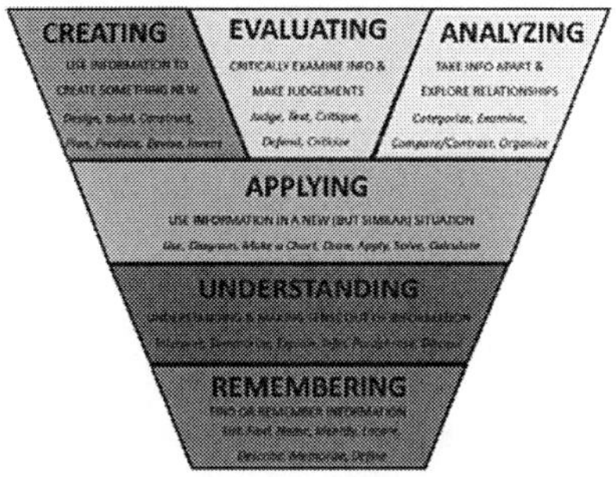

Bloom's Taxonomy

Assessment in a Virtual Environment is both **implied** and **overt**

I.e., if a learner is prompted to "Call the Flight Deck from ML3 Station", if the user can successfully perform this task, we know:

- The user can <u>locate</u> ML3
- The user can <u>locate</u> ML3 Station
- The user can <u>locate</u> the cabin interphone
- The user can <u>identify</u> the purpose of the Cabin Interphone
- The user can <u>identify</u> how to use the buttons on the interphone
- The use can correctly <u>perform</u> a phone call to the Flight Deck using the interphone

CUBIC

Cubic Proprietary

Zone of Proximal Development and Progressive Game Difficulty

40

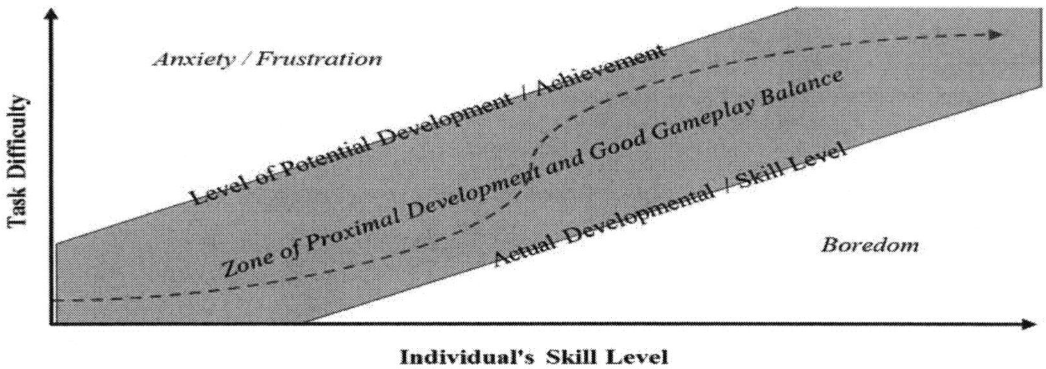

CUBIC

Cubic Proprietary

41

Scaffolding Instruction

Techniques:

- Levels of "hints"
- Practice levels (training wheels)
- Non-player character help
- Tutorial walk-throughs
- Use judiciously

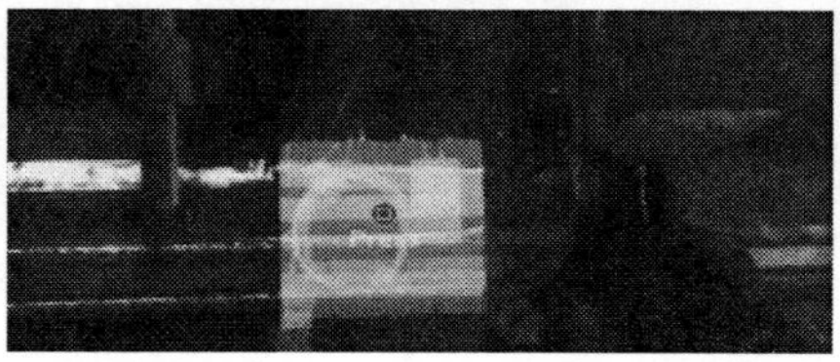

CUBIC

Cubic Proprietary

42

Feedback

- Deliberate practice requires feedback for improvement
- Feedback types
  - Naturalistic – realistic, true-to-life
  - Instructional – warnings, cautions, messages, scores, hints
- Both important, but they should be carefully constructed
- Balance between not enough and too little
  - Sometimes it's better to stop someone, sometimes it's better to continue on the lesson

CUBIC

Cubic Proprietary

Design Based on Analysis

Identify Training System Requirements

Cost-Benefit Analysis (if required)

Identify Game Characteristics (play, story, etc.)

Organize Instructional Content

Sequence Instruction

Identify Learning Strategies/ Prescriptions

Generate Learning Objectives/

Identify Administrative Training Needs

CUBIC

Designing the immersive virtual environment

Platform Design

- What kind of technology will you use?
  - What kind of technology is best suited to the training?
- Deployment strategies
  - Mobile
    - Phone vs. Tablet
    - iOS or Android or Windows
  - Online or offline
  - Live
  - Synchronous/Asynchronous
- Impacts: game engine/3D modeling, scalability, lesson/content size, design, input/output devices…

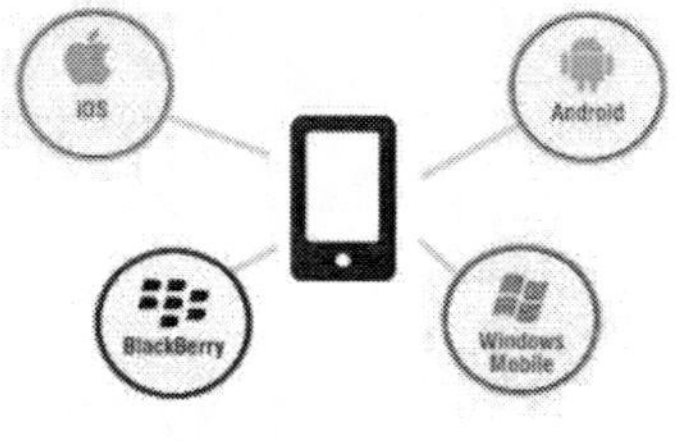

Fidelity Design

- What is fidelity?
  - Measure of **realism** for any representation, or model, of the real world
  - Fidelity has a linear relationship with cost
  - **Diminishing returns on fidelity cost!**
- Types of Fidelity:
  - Functional Fidelity
  - Psychological-Cognitive Fidelity
  - Task Fidelity
  - Physical Fidelity
    - Visual-Audio Fidelity
    - Motion Fidelity
    - Equipment Fidelity

47

Fidelity Design

- Functional Fidelity:
  - Degree to which a device replicates the actual environment; how real the simulation appears and feels
  - Parameters: time, role accountability & responsibility, environment behavior and consequence, reactions to tasks and inputs by the learner
- High functional fidelity doesn't necessarily mean high physical fidelity
  - E.g., Microsoft Xplane with a game controller

CUBIC

48

Fidelity Design

- Psychological-Cognitive Fidelity
  - Degree to which the simulation and associated devices replicate the cognitive and psychological factors
  - Parameters: stress, fatigue, workload, noise, heat
- Psychological-cognitive fidelity does not require true-to-life stimuli…
  - Show a timer
  - Use a haptic vest
  - Noise
  - Ice buckets
  - Lab coats

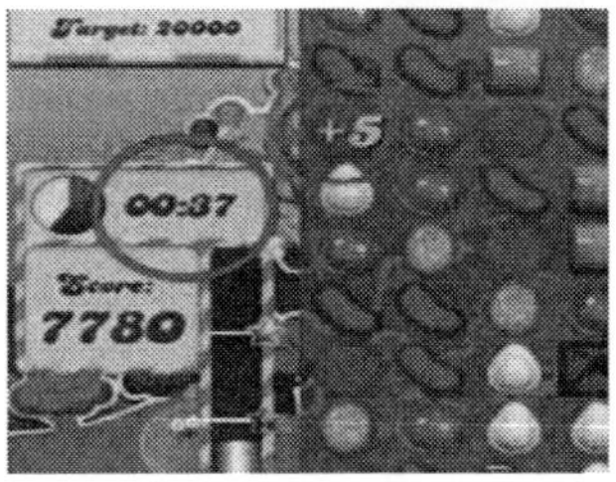

CUBIC

Fidelity Design

49

- Task Fidelity
  - Replicating tasks, procedures, and maneuvers the learner performs
  - Parameters: ensuring steps and required elements are true to life checklists, using the right equipment, communications, traversing space, etc.
- Task fidelity does not require any other components
  - Role playing has high task fidelity (live and games)

CUBIC

Cubic Proprietary

Fidelity Design

50

- Physical Fidelity
  - Degree to which the device looks, feels, and sounds like the actual environment
  - Parameters: The actual environment visual displays, controls, and audio as well as the physics models
  - **Visual-Audio Fidelity**
    - Replicating the visual and auditory environment
  - **Kinesthetic/Haptic Fidelity**
    - Replicating the motion or tactile cues in the live environment
  - **Equipment Fidelity**
    - Replicating the actual equipment via hardware, software, or both

CUBIC

Cubic Proprietary

Fidelity Design

51

- Approximate some Kinesthetic & Equipment Fidelity using a Virtual Environment
  - See the hardware move, how to use it
  - See the physicality of specific tasks

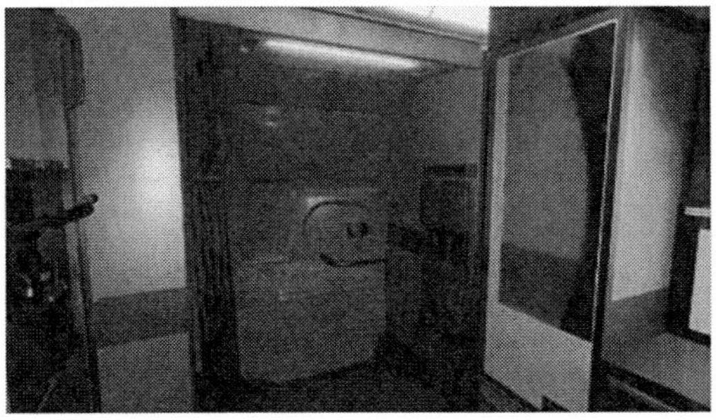

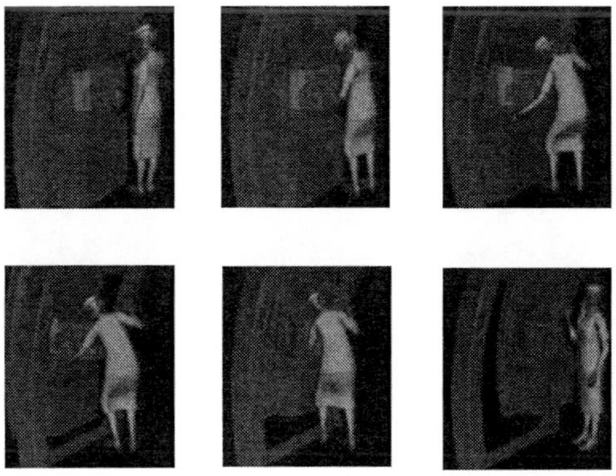

CUBIC

Cubic Proprietary

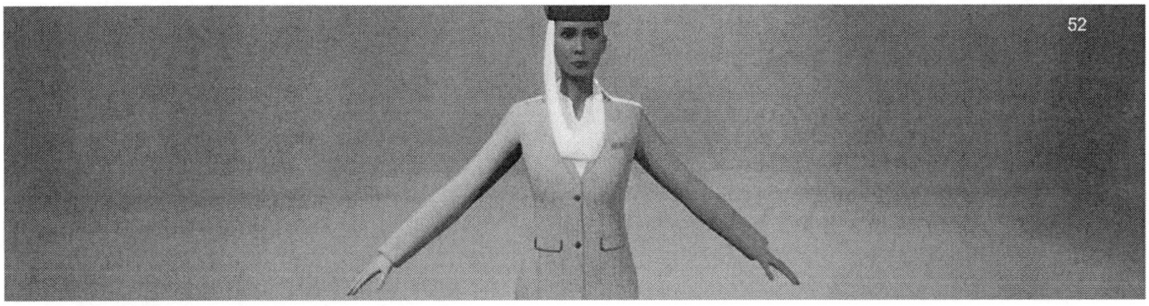

52

Character Avatars

- Use as mentors
- Use as character avatar instrumental to training
- Use as non-player characters

Considerations:

- Are there image standards?
- Are there branding considerations?
- How likely would this be to change?

CUBIC

Cubic Proprietary

53

Why all the details?

- Intuition is really experience in disguise.
- Schemas:
  - Match real-world expectations
  - If something doesn't fit, ***why***?

Decision-makers ask "what's going on here?" before they ask "what do I do?"

CUBIC

Cubic Proprietary

54

Special Considerations

- Wearables
  - Wearables work well for 3D rooms, but limits deployment
  - Physical considerations of trainees
- Simulator Sickness
  - Can happen in those not used to immersive environments
  - Vast majority of effects can be overcome by good design
    - Feedback, animation, navigation, etc.
- Accessibility
  - What are the considerations of the population for accessibility via language, learning disability, physical disability, etc.?

CUBIC

Cubic Proprietary

55

Gamification Strategies

Cubic Proprietary

56

Gamification

- Not all games are simulations, and not all simulations are games
- What makes a game?
  - Story
  - Flow
  - Motivation
    - Individual Intrinsic
    - Individual Extrinsic
    - Social

CUBIC

Cubic Proprietary

Why Not Stick with Simulation?

57

- Games support and improve training outcomes
  - People who have game-based training play to train, AND play again for fun
  - Presents information as part of a meaningful, scaffolded experience
  - Games naturally support other reward mechanisms for performance: awards, leader boards, etc.

- **Simulations** don't provide corrective feedback
- **Training** can be boring, and can limit learning
- **Games** are fun and motivating and memorable, but do not provide adequate instruction

Game-based training in Immersive Environments is the best of all three worlds

CUBIC

Cubic Proprietary

Flow

58

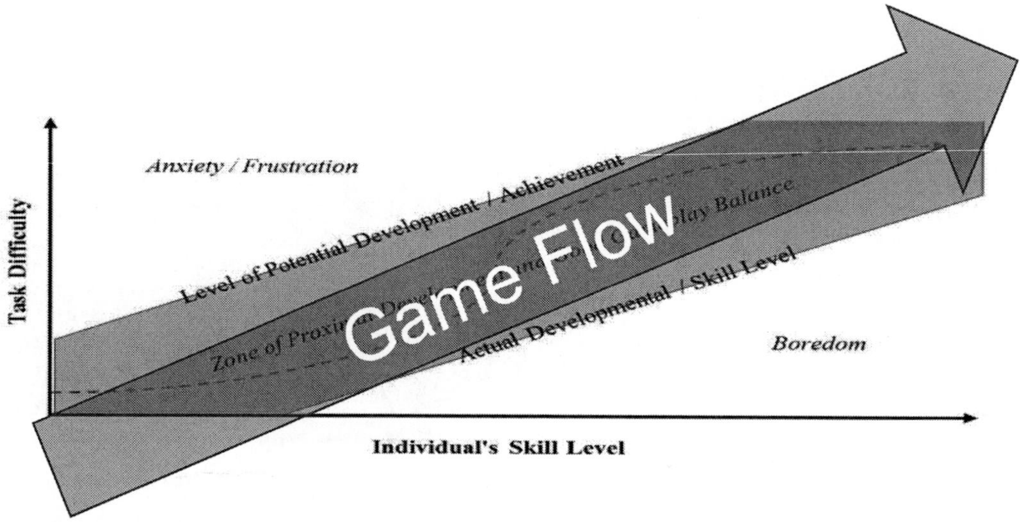

CUBIC

Cubic Proprietary

Gamification Strategies

59

- Games should go from easy to hard, like training
- Start with fundamentals, then add more scenarios
- Build in difficulty and nuance

- E.g.,: For Cabin Crew, Ditch = Crew Systems + Doors Arming/Disarming + Door Opening + Door Emergency Assessment + Slides + Evacuation + Slide/Raft Disconnect + Survival + Medical

- https://www.youtube.com/watch?v=mCis7kkOrTk

CUBIC

Cubic Proprietary

Goals

60

- Providing meaningful benchmarking goals tied to performance
  - Specific goals are the best
  - Directs attention on the task
- Consistency

- Difficult goals = increase in effort and better performance

- Stimulate prior knowledge and skill

- Satisfying!

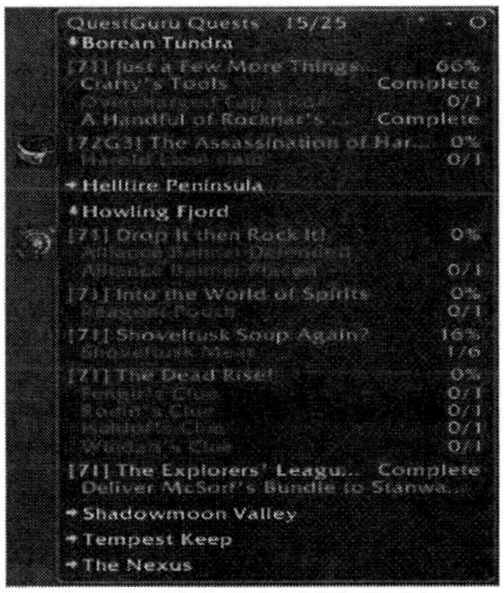

CUBIC

Cubic Proprietary

Mini-Games

61

- Use mini-games as rewards mechanisms
  - Fun
  - Add rewards/bonuses
- But still provide meaningful training
- Way to mix types of games within a single game
  - E.g., add a casual game to an immersive game
  - Target multiple audiences

CUBIC

Cubic Proprietary

Game Types & Game Play

62

Example Game Genres for Specific Learning Outcomes		
Learning Outcome	**Game Genre**	**Rationale**
Skill-and-drill	Fighting /First person shooter	Relies on patterns and repetition; player initiates action.
Declarative knowledge, facts, statements	Classic arcade-style, puzzle	Simpler gameplay, makes use of both pattern matching and recall abilities.
Scientific concepts, problem solving	Action-Adventure	Objective-driven nature and use of puzzles promote creative problem-solving. Problem-solving helps players create an understand connections between concepts (Spires, Rowe, Mott, & Lester, 2011).
Complex relationships, decision-making	Real-time strategy (RTS)	Gameplay relies on resource management and complex system reactions, which provides analogous practice for this outcome.
Metacognitive (dynamic, complex) knowledge	Traditional and sandbox role-playing games (RPGs)	Requires planning and forethought at the crux of gameplay. Branching narratives and interactions with non-player characters (NPCs) with quest systems can encourage deeper processing and teach more complex topics.

CUBIC

Cubic Proprietary

63

Individual Intrinsic & Extrinsic Motivation

- **Intrinsic Motivation**
- Individual motivational factors based on values
- What are your learners' values?
  - Pride, company loyalty, feelings of self-value

- **Extrinsic Motivation**
- Doing something for external gain

Cubic Proprietary

64

Individual Intrinsic Motivation Examples

- Scoring
  - Stars, points, etc.
- Badges
  - Reward incremental process
  - Reward good training behavior
  - Think: 2 stars vs. 3 stars
- Sense of Goal achievement (job well done)

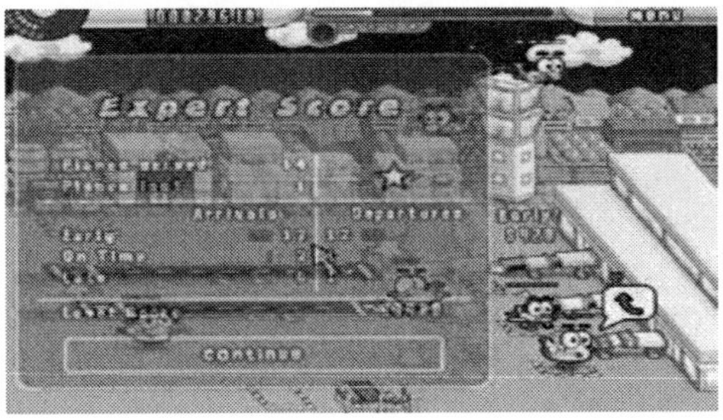

Cubic Proprietary

Individual Extrinsic Motivation Examples

65

- Rewards tied to points/success metrics
  - Bonuses, prizes, etc.
- Leveling up
- Bonuses, freebies, equipment that helps the gameplay
  - Make sure this ties to the learning!
- Content
  - Mini games, choose your sound track, "hidden" levels, etc.

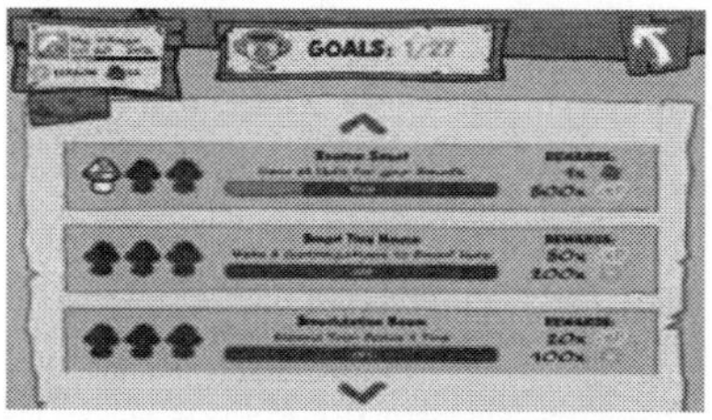

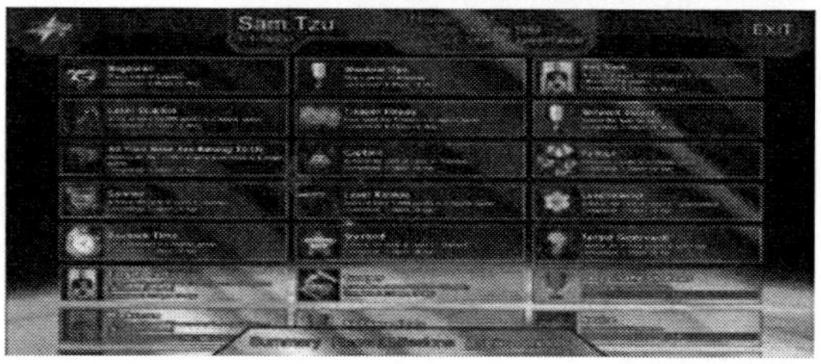

CUBIC

Cubic Proprietary

Social Motivation

66

- Leaderboards
- Gameplay with others of that level
  - Relevant in multiplayer
- "Share" to get bragging rights
- Competition factor

Dr. Sheldon Cooper, character on the CBS show *The Big Bang Theory*, who is fiercely competitive and delights in besting others, saying "Bazinga!" when he has.

CUBIC

Cubic Proprietary

Use Multiple Motivational Factors

67

- Kaplan University found motivations for non-traditional learners based on their unique factors rewarding good learning behavior
  - raised the time students spent in class by 17%
  - increased the percentage of students who chose more challenging coursework by 85%
  - raised the prevalence of higher grades by 9%
  - Badges, leaderboards: Teamwork, Dedication, Problem-Solving, etc.

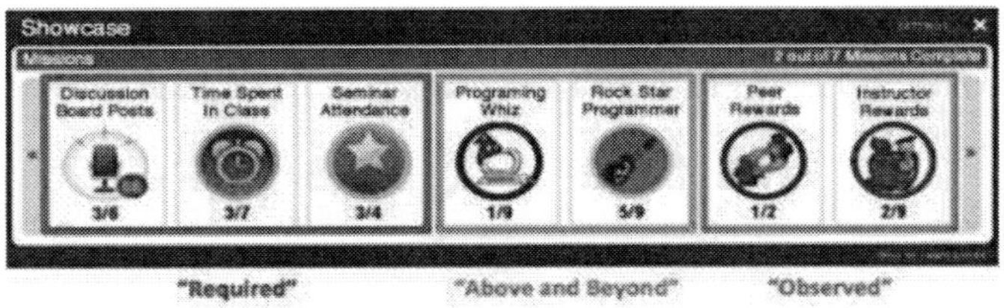

CUBIC

Cubic Proprietary

Wrapup

68

- Game-based training is a powerful and cost-effective training tool when designed well from the ground up
- Solutions aren't one-size fits all and should be custom for your organization
- First, determine what you need to train
- Next, design the training itself – content and instructional strategy
- Then, design the virtual environment to provide the critical cues you need and deploy on the best devices
- Finally, design gamification strategies to complement the learning
- But always center on the learner

CUBIC

Cubic Proprietary

69

Thank you!

Questions? Comments? Follow-up?

Email

Amanda Bond at amanda.bond@cubic.com

or email

Kellie White, Training Manager, Safety and Emergency Procedures, Emirates at kellie.white@emirates.com

CUBIC

Cubic Proprietary

Trends and Outlook
in
Projection Technology

Dave Fluegeman, VP Simulation, Barco
Dave.Fluegeman@barco.com

Barco Simulation
Engineered to Outperform

BARCO
Visibly yours

Looking Back

- Increased fidelity
- Higher performance
- The COTS Revolution

Barco Simulation
Engineered to Outperform

BARCO
Visibly yours

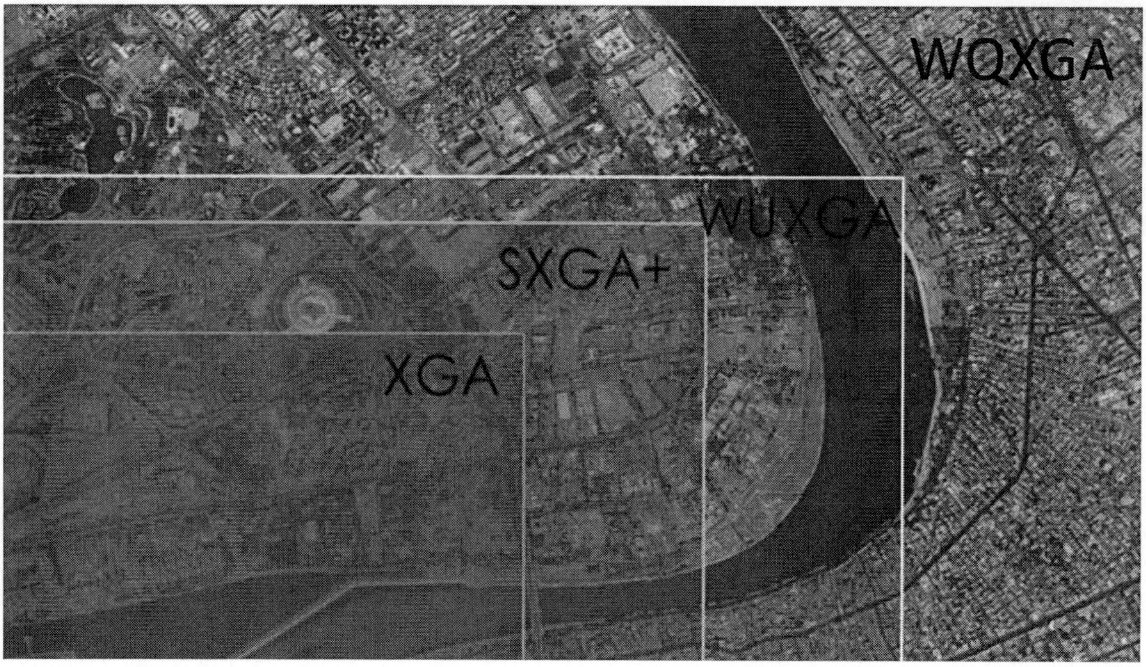

Display Considerations

- The Continuing Debate of "Good Enough"
  - Something not present in consumer markets
- Contrast, Color, and Brightness
  - LAMPS ARE DEAD!!!!
  - Or, Are They????
- Display performance over time and contribution to higher rates of availability

Barco Simulation
Engineered to Outperform

Illumination Technology Choices

- Illumination sources:
  - Lamps
  - LED
  - Laser

LED and Laser } Solid State Illumination (SSI)

Barco Simulation
Engineered to Outperform

BARCO visibly yours

Total Cost of Ownership explained

- **Lamp:**

17h a day x 300 days = 5100h x 10 years = 51000h/2000h ~ 26 lamp changes

+ \$400 x 3 lamps (cost of lamp):	= \$31,200
+ \$200 (for 2 hours) maintenance cost	= \$ 5,200
+ lost revenue (downtime) \$500 x 2h	= \$26,000
Typical maintenance cost 10 years	**= \$62,400**

Barco Simulation
Engineered to Outperform

BARCO visibly yours

Trending: 4K resolution

- Benefits of pixel-shift
- Native 4K vs. 4K UHD

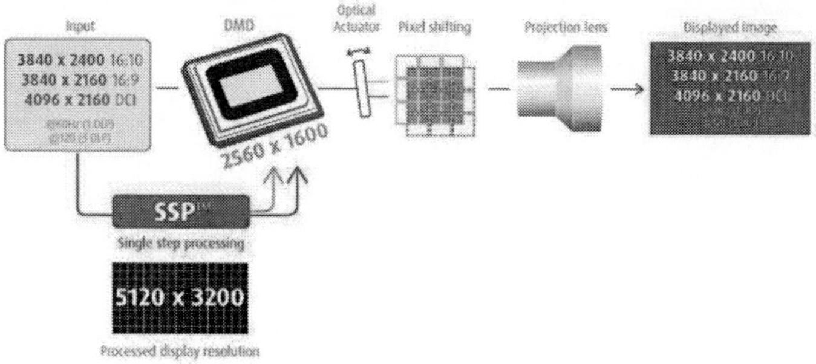

One example of Pixel Shift using DLP technology
Shown here Single Step Processing (SSP™) by Barco

Barco Simulation
Engineered to Outperform

BARCO
Visibly yours

Continued Evolution of Display Technology

- Evolution of LED Illumination
  - Can They Keep Pace?
- Direct Laser
- Flex LED/OLED

Barco Simulation
Engineered to Outperform

BARCO
Visibly yours

wats2016 ORLANDO
ROSEN SHINGLE CREEK
ORLANDO, FLORIDA
APRIL 19-21 2016

The Benefits of Glass Mirrors on Flight Simulators

Justin Knaplund, Senior Staff Display Systems Engineer,
FlightSafety International, Austin, Texas, USA

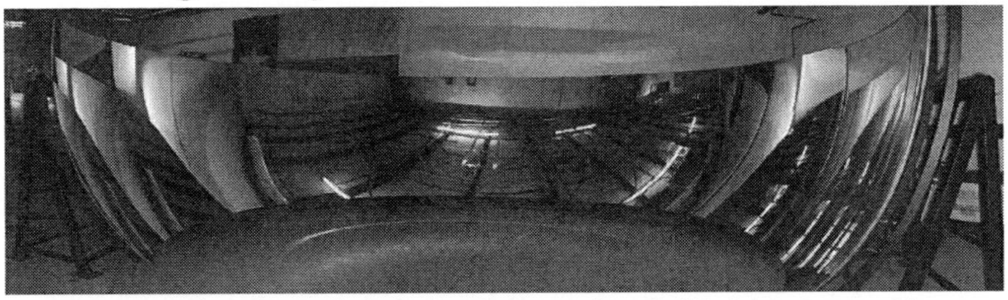

Presentation Outline

- Collimated displays – how & why
- Types and limitations of large collimated displays
- The trend towards increasing field of view for training
- Case studies of large field of view systems that have been fielded

Collimation – the process of making light rays parallel so they appear to be at or near infinity

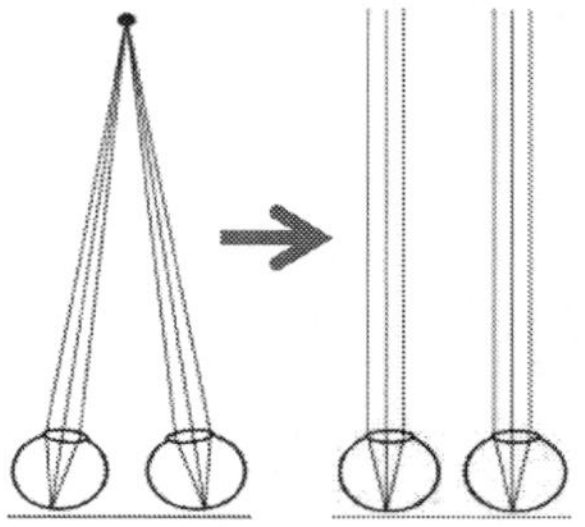

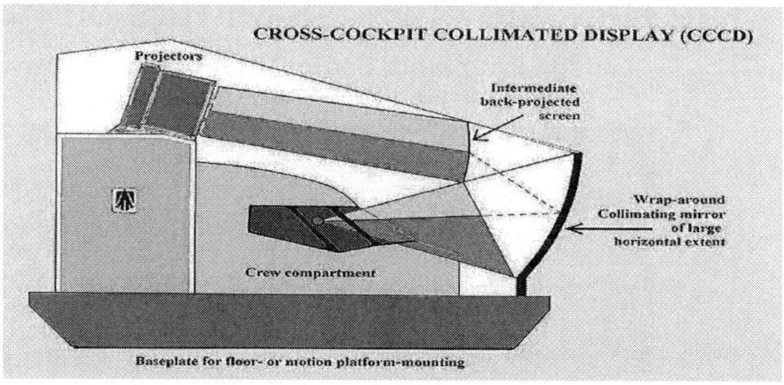

Courtesy of Wikipedia; https://en.wikipedia.org/wiki/Collimated_light

The main benefit of a collimated display - realism

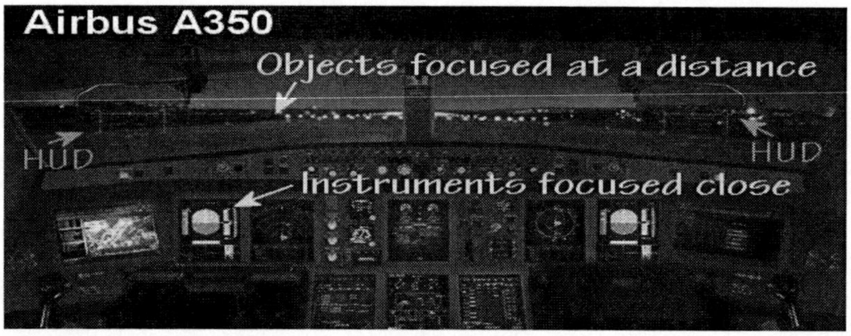

1. Eyes must refocus
2. Heads Up Displays
3. Helmet mounted displays
4. Cross-cockpit view the same from both seats

Glass versus Mylar Mirrors - Differences

Characteristic	Glass	Film
Mirror Shape/Distortions/Roll-Off	√	
Field of View	√	
Impact Durability	√	
Trim Cuts	√	
Cleaning	√	
Weight		√
Seams		√

FlightSafety international *"The best safety device in any aircraft is a well-trained crew."* Slide 5

Glass versus Mylar - Distortion

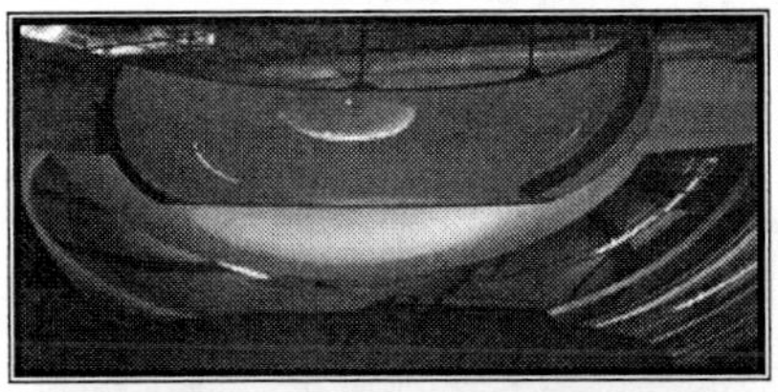

<u>Glass</u>:
½" thick substrate
Ground, polished and coated
Radius tolerance 0.5% of radius
Mirrors produced in radius-matched sets (within 0.5% of each other)
"Blur circle" = 1" diameter or less

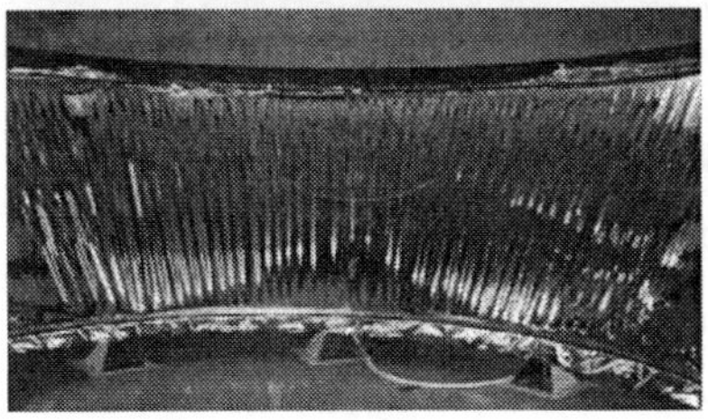

<u>Mylar</u> (shown before vacuum):
Aluminized Polyester Film
Shaped by stretching over a hollow mirror cell with film clamped or taped around edges and applying vacuum.
Tension in the mirror film is higher vertically than horizontally
Horizontal tension is higher in center than along the top/bottom

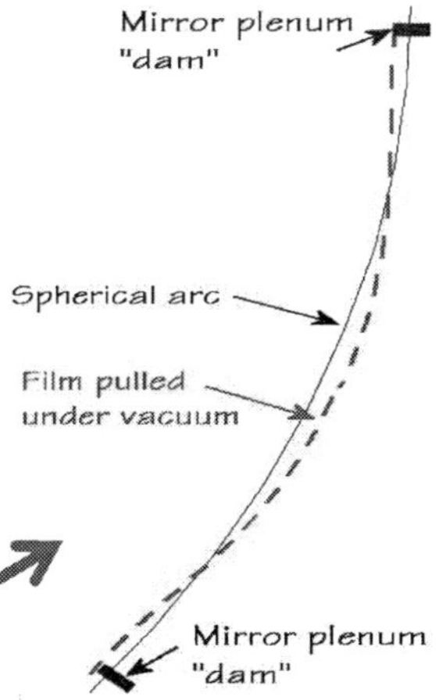

FlightSafety international *"The best safety device in any aircraft is a well-trained crew."* Slide 6

Image Roll Off around edges of Mylar Mirror

Mylar distortion issues: Ground rush, height errors, air refueling in distortion area, and peripheral "star field acceleration" at night

FSEMC 2008 FedEx Presentation "Accommodating an Aircraft Head Up Display in a Wide type Visual Display System" showed erratic glideslope as a result of Vertical Compression found in Mylar mirror

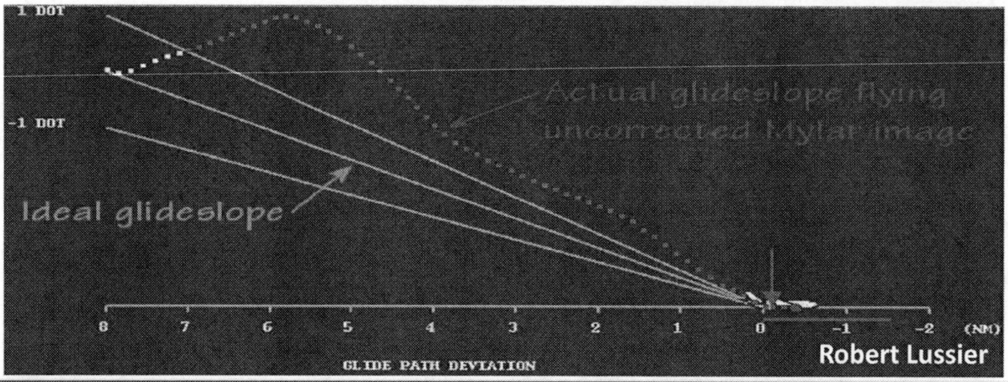

Glass Mirrors do not suffer edge distortion

FlightSafety international "The best safety device in any aircraft is a well-trained crew." Slide 9

Horizontal Field of View (FOV)

Mylar film limits:

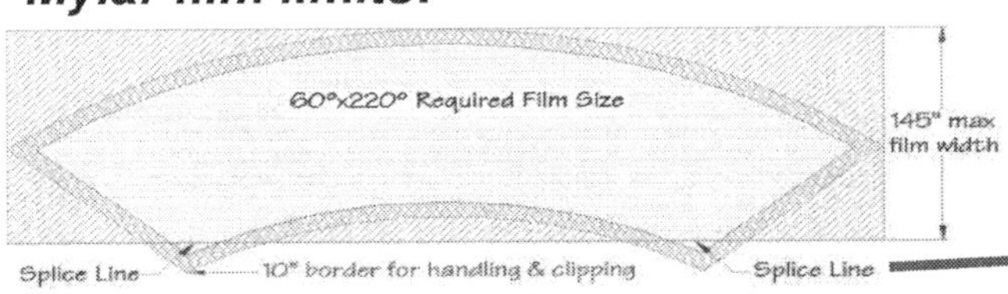

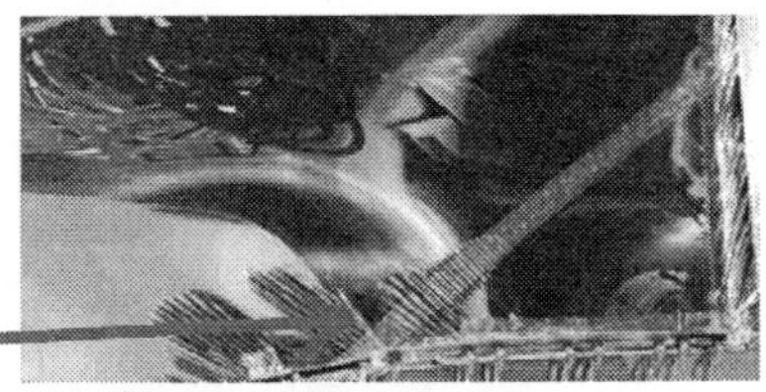

Film mirror FOV is limited by polyester film width; FOVs > ~60°x225° require "splicing"

Glass mirrors:

Glass mirror can have as many segments as needed to meet horizontal FOV.

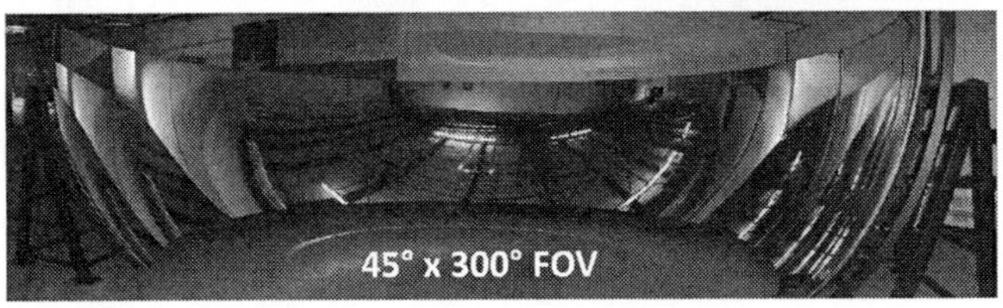

FlightSafety international "The best safety device in any aircraft is a well-trained crew." Slide 10

Why more Horizontal FOV?

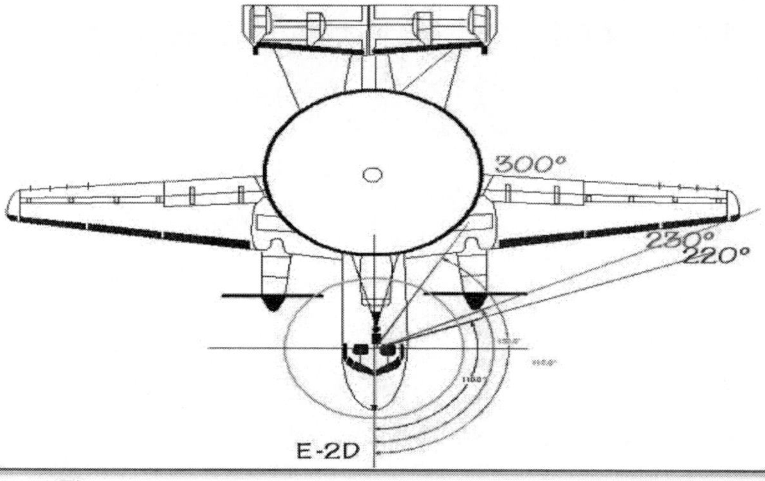

Increased Horizontal FOV Benefits:

- **Additional simulator training tasks - engine out procedures, etc**
- **Enhanced "immersive" realism**
- **More peripheral cues for speed, attitude, etc = better training**
- **Enemy aircraft threat assessment**
- **Overhead landing patterns**

FlightSafety international *"The best safety device in any aircraft is a well-trained crew."* **Slide 11**

Horizontal FOV video: E-2D 300°x45°

FlightSafety international *"The best safety device in any aircraft is a well-trained crew."* **Slide 12**

Horizontal FOV 220° vs. 290°

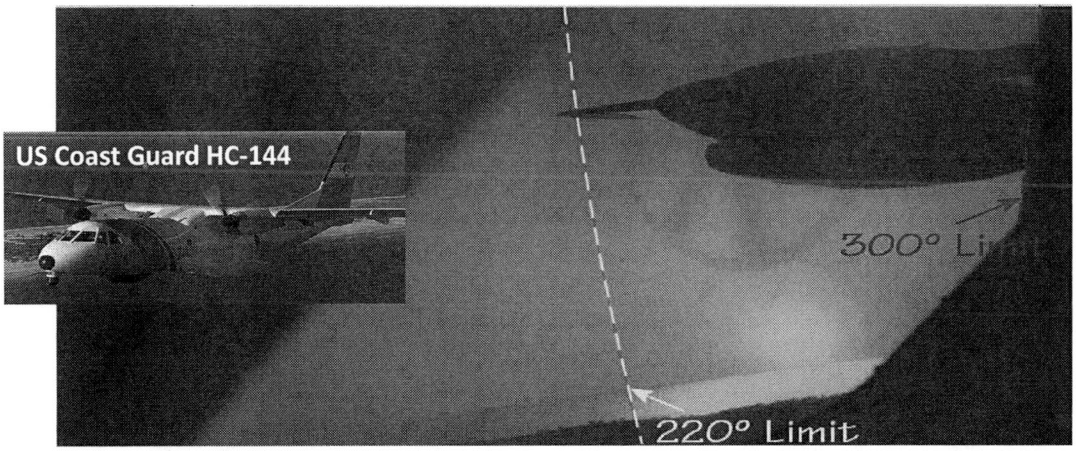

HC-144, 290°x60° assembly video

Why more Vertical FOV?

Increased Vertical FOV helps with:

- Take-off & hover cues
- Precision approaches to confined areas
- Obstacle avoidance
- Precision spot landings
- "Brown Out" landing training
- Uplook for undistorted air-refueling

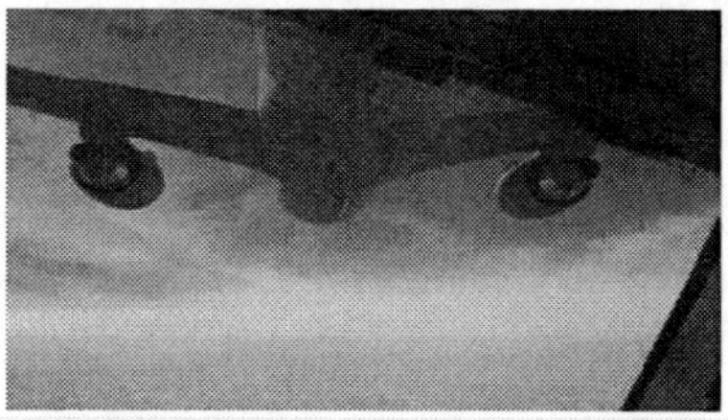

FlightSafety international *"The best safety device in any aircraft is a well-trained crew."* **Slide 15**

Ever increasing military FOV requirements

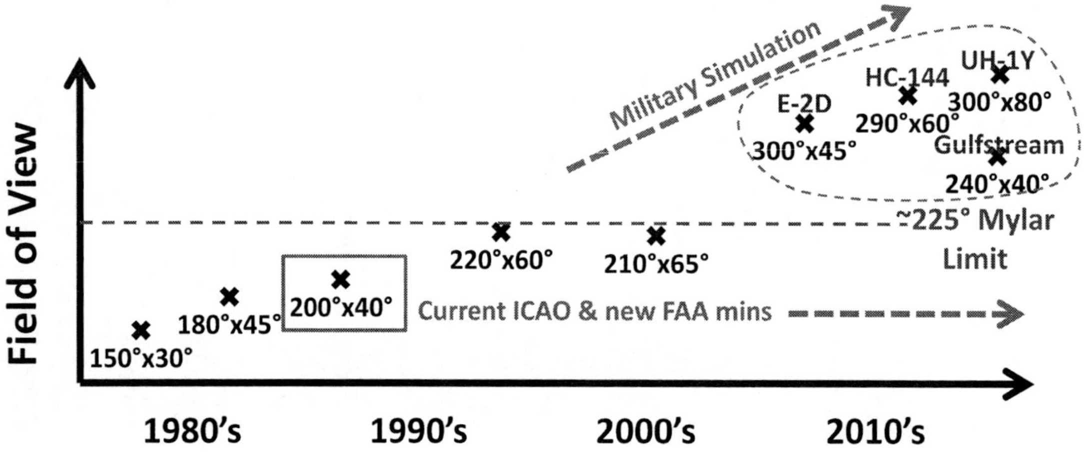

FlightSafety international *"The best safety device in any aircraft is a well-trained crew."* **Slide 16**

Glass Mirror Durability (with proper backing)

Test of 8" Phillips screwdriver drop on uncoated glass mirror (VIDEO)

Mirror impact testing result, 4' drop

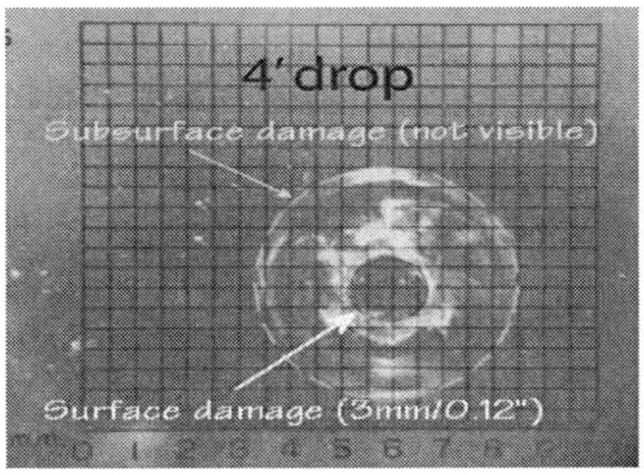

FlightSafety international

Case study #1: US Navy E-2C/D, 2009

- US Navy wanted to move training tasks like threat recognition & prop feathering to the simulator, resulting in a requirement for a 300°FOV display
- At the time, no collimated system had ever been developed to cover > 225°
- Multiple glass mirror segments (7) allowed a full 300°x45° FOV coverage
- Program success led to industry acceptance of large FOV systems

FlightSafety international *"The best safety device in any aircraft is a well-trained crew."* **Slide 18**

Case study #2: US Coast Guard HC-144, 2013

- Task 1: Viewing the runway during extended downwind pattern so that the majority of transition training can be performed in the simulator
- Task 2: Providing visual feedback to the crew in case of engine fire or smoke
- Task 3: Viewing the leading edge of the wing for ice formation and visually monitoring the effective operation of the de-icing boots
- Task 4: Observing the nacelle/spinner assembly to determine if the propeller is feathered completely or windmilling, as this directly affects drag and the subsequent checklist to follow. This training task cannot be safely performed in the aircraft

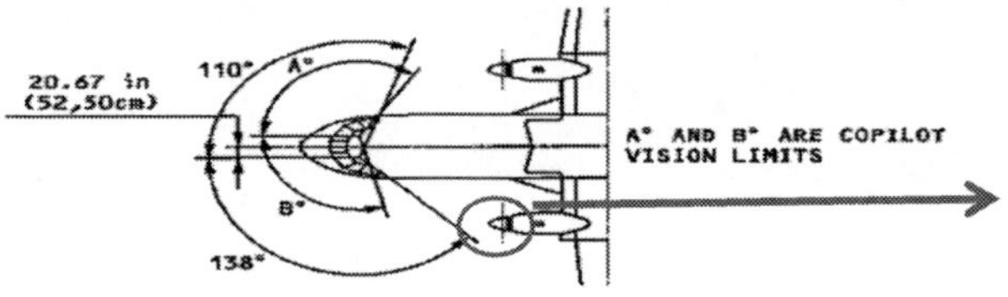

FlightSafety international "The best safety device in any aircraft is a well-trained crew." Slide 19

Case study #3: USMC UH-1Y, 2014

- In 2011, a Bell Helicopter study determined that the current standard 220°x60° FOV display was unacceptable for adequate training in such tasks as: Hover, taxi, take-off and landing, and maintaining visual contact with threats or wingmen
- The study concluded the FOV should be 300°H x 60°V collimated with side & chin displays that increase the vertical FOV up to 80°V total

1st UH-1Y system (fore-ground) & 2nd system (background) during assembly at the FlightSafety facility in Broken Arrow, Oklahoma, in 2014

FlightSafety international "The best safety device in any aircraft is a well-trained crew." Slide 20

Case study #4: USMC CH-53K, 2015

• Field of view = 260°H x 60°V collimated "Out the Window" with auxiliary side & chin displays that increase the ***vertical FOV >80°V total***
• Visual display must be installed ***inside a relocatable container***

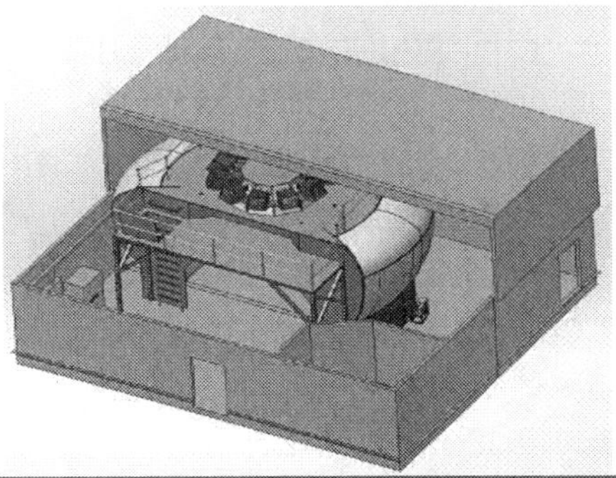

Case study #5: Gulfstream Immersive Flight Simulation Lab, 2014

• Gulfstream engineers performed a series of pilot tests with various FOV systems (using flat monitors) and determined that >225° horizontal FOV was required for pilots to quickly recognize & recover from turbulent input
• The FOV requirement was 240° x 40° and had to fit inside the 14' lab height

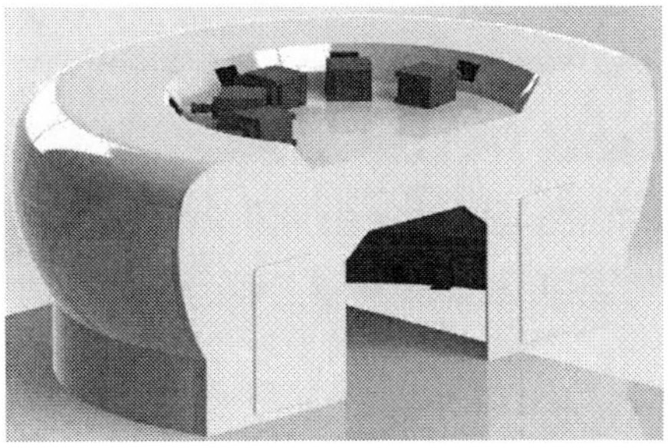
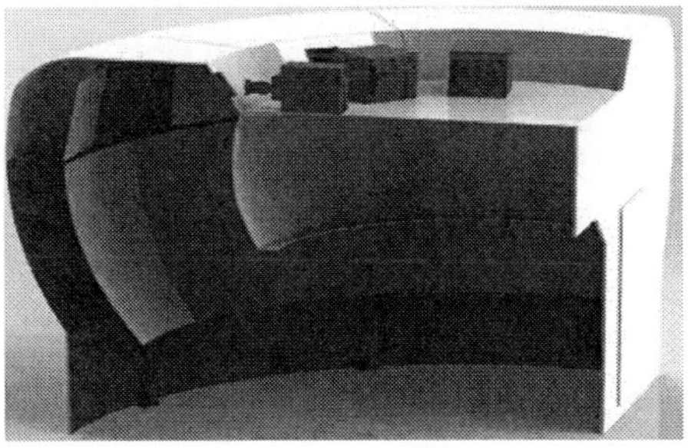

The Benefits of Glass Mirrors on Simulators Summary:

- The trend in flight simulation is for increasing Field of View (FOV) in order to perform more training tasks in the simulator:
  - Increased horizontal FOV enhances realism, allows a view of critical components such as engines, and permits overhead traffic pattern practice.
  - Increased vertical FOV improves take-off, landing and air-refueling tasks. Beyond 60° requires auxiliary displays.
- Only glass mirror displays allow for horizontal FOV's >>225° and for a continuous vertical FOV between mirror and auxiliary screens.
- Glass mirrors also have superior optical qualities compared to Mylar film, resulting in much less image distortion throughout the display.

FlightSafety international *"The best safety device in any aircraft is a well-trained crew."* Slide 23

Questions?

Justin Knaplund, Senior Staff Display Engineer, FlightSafety International, Austin, TX. justin.knaplund@flightsafety.com

Don Conklin, Director of Business Development, FlightSafety International, Austin, TX. donald.conklin@flightsafety.com

FlightSafety international *"The best safety device in any aircraft is a well-trained crew."* Slide 24

MOOG

Changes in 2nd Generation Level D EM Technology

April 21, 2016

MOOG

Agenda

- History of Electric Motion
  - First EM System
  - First Generation EM Technology
    - Market Needs
- Market Dynamics
  - Impact of 2008 Financial Crisis
  - Reduce Cost of Ownership
- Second Generation Electric Technology
  - Improved Reliability
  - Easier to Use
  - Easier to Maintain
  - Improved Spares Availability
  - Environmentally Friendly

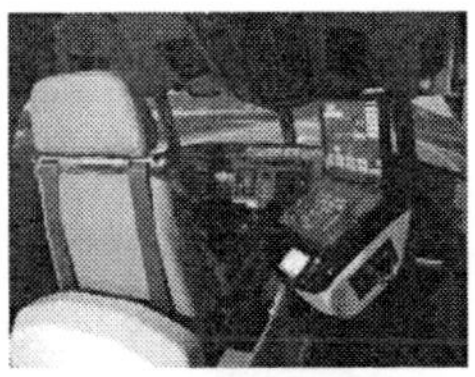

History Of Electric Motion

- The first EM motion system attained level D status in '02.
  - EM faced a very skeptical market.
  - Adaptation rate was thought to be very gradual.
  - Emphasis was on:
    - Meeting level D criteria.
    - Safety
    - Providing the same level of safety as hydraulic systems.
      - Return to Home
      - Runaway prevention/reduction

Electric Motion History

- EM has come a long way in a short time:
  - Adaptation was much faster than originally anticipated.
  - Original production rates of EM systems grew at an annual rate of 86% over the first 3 years.
  - Currently, there are over 500 EM systems currently rated for level D.
  - Safety and level D performance are now threshold criteria.
    - Today's customers demand a lower cost of ownership

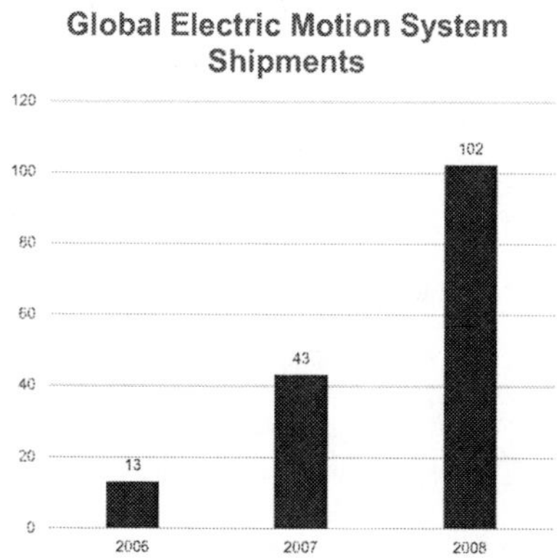

- Source: Moog Internal shipment data and 2015 CAT Civil & Mil. Flight Simulator Census Data

Market Dynamics-2008 Financial Crisis

- The financial crisis of 2008 had a major impact on Commercial Airlines.
- This impact trickled down through the entire industry.
  - Flight Training Centers
  - Manufacturers of Full Flight Simulators
  - Manufacturers of electric motion systems
- The entire industry demanded a lower Cost Of Ownership
- 2nd Generation EM Motion systems have:
  - Greater Reliability
  - Easier to use
  - Easier to Maintain
  - Environmentally friendly
  - Systems that provide greater flexibility

Year Over Year Revenue per KM

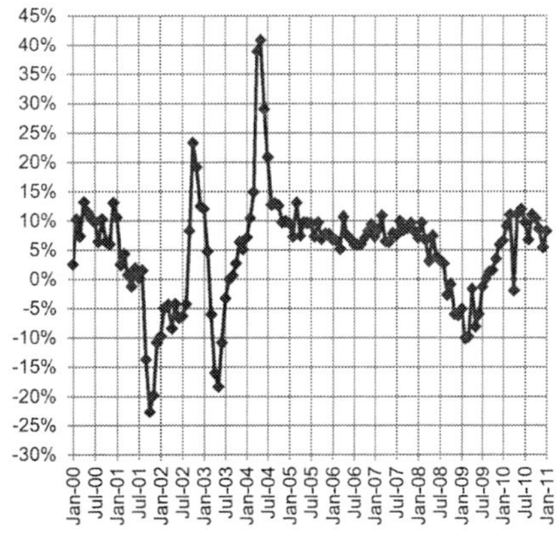

Source: International Air Transport Association

2nd Generation EM Technology

Improved Reliability

- Digital Feedback devices
  - More robust
  - Less vulnerable to electrical noise
- Safety PLC eliminates relays and PLC boards
  - Solid State PLC
  - Reduction in parts
- Elimination of Nitrogen in mechanical safety circuit
  - Reduces chances of leakage.
- Greater reliability through the introduction of Statistical Process Control into the manufacturing process.
  - Improvement in overall reliability through improved testing and characterization of critical performance criteria.

2nd Generation EM Technology

MOOG

Ease Of Use

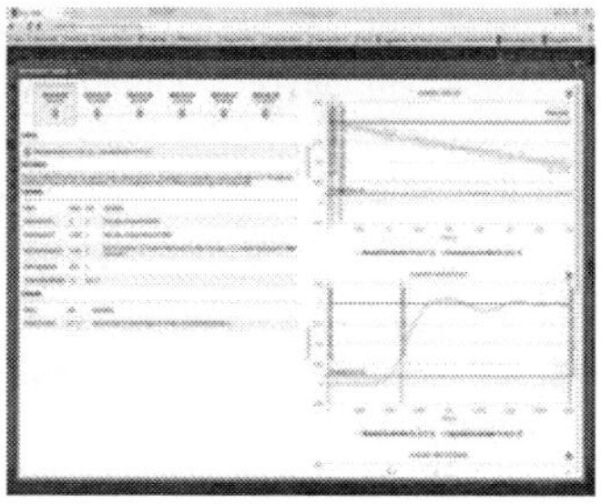

- Improved Software
  - Easy to use Graphical User Interface
  - Connection through company intranet eliminates need for mouse or keyboard.
  - AOG diagnostic files can be e-mailed from web interface.
- Automated Test Tool
  - Provides automated method of testing motion to Level D specifications.
  - Reduces test time from days to minutes.
  - Delivers pre packaged reports on performance vs. Level D
  - Establishes baseline criteria to establish the impact of changes

2nd Generation EM Technology

MOOG

Improved Availability

- Deterministic diagnostics
  - Reduces the need for trouble shooting during AOG
  - Digital busses provide more diagnostic information.
  - Diagnostic information on all devices including circuit breakers and contactors
- Global Service and Spare Capabilities
  - 21 day turn around on spares.
  - Rotable Spare Inventory
  - Local Spares Depots
  - Local Service Capability
    - Same time zone
    - Same language

2nd Generation EM Technology

MOOG

Improved Performance

- Reduction in turn around bump.
  - New software algorithm reduces turn around bump relative to 1st generation EM systems
- Reduction in audible noise.
  - New nut design on screw reduces audible noise.

*Used with permission of FlightSafety International

Conclusion

MOOG

The changes incorporated in 2nd generation electric motion systems are not all evident to the eye.

- Many of changes are hidden to the user.
  - Software
  - Services
  - Diagnostics
  - GUIs
  - Ball Screws

The benefits associated with 2nd generation electric motion systems reduce the cost of ownership for OEM and End User customers.

Question & Answer

MOOG

The Next Generation of Aviation Training
Professor Lori J. Brown
Western Michigan University
College of Aviation
ICAO Next Generation of Aviation Professionals
(NGAP) Outreach Chair

The NEW Generation of Aviation Training
• Augmented reality
• 3D Modeling
• Cost effective solutions

Alpha Generation

Français
INTERNATIONAL CIVIL AVIATION ORGANIZATION
A United Nations Specialized Agency
Search this site...
Sign up for updates
About ICAO | Strategic Objectives | Aviation Development | Events | Publications | Training | Store | Careers
Next Generation of Aviation Professionals
Home
About NGAP
NGAP Programme
States & Organizations Supporting NGAP
Aviation Tertiary Education Directory (ATED)
Presentations
Videos
Internship Opportunities
Events
1st NGAP Symposium
2nd NGAP Symposium
ICAO > Safety > Next Generation of Aviation Professionals
Next Generation of Aviation Professionals (NGAP)
It has been recognized by the international aviation community that there will be an anticipated shortage of skilled aviation professionals in the near future. In order to address this important issue, ICAO launched the Next Generation of Aviation Professionals (NGAP) initiative to ensure that enough qualified and competent aviation professionals are available to operate, manage and maintain the future international air transport system.

Organizations which have supported ICAO NGAP

States / Regulatory Authorities

International Organizations

Education / Training Providers

Industry/Others

The Future is about Seeing More

Augmented reality

NLR augmented reality concept demonstrator. www.nlr.hl

The Future is about Feeling more with Haptic Technology

- Sound travels through air by creating a pressure differential
- UltraHaptics use a small collection of ultrasonic speakers, concentrating the sound waves to a specific point

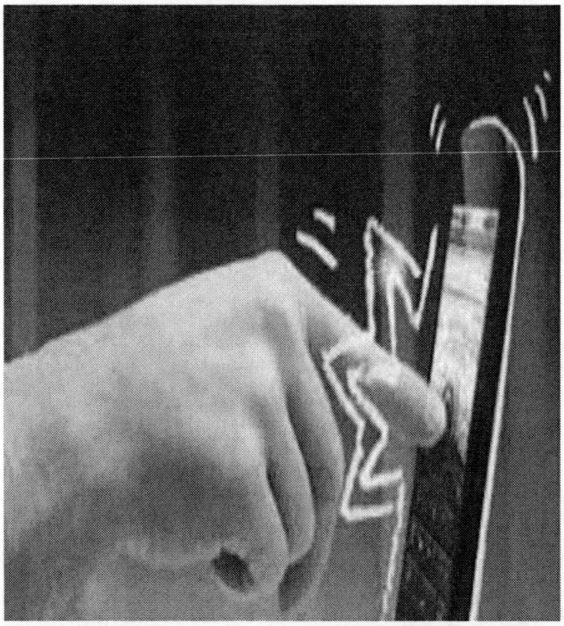

Further Development

- Virtual Reality Glasses

- Augment Reality Glasses
- Hapti

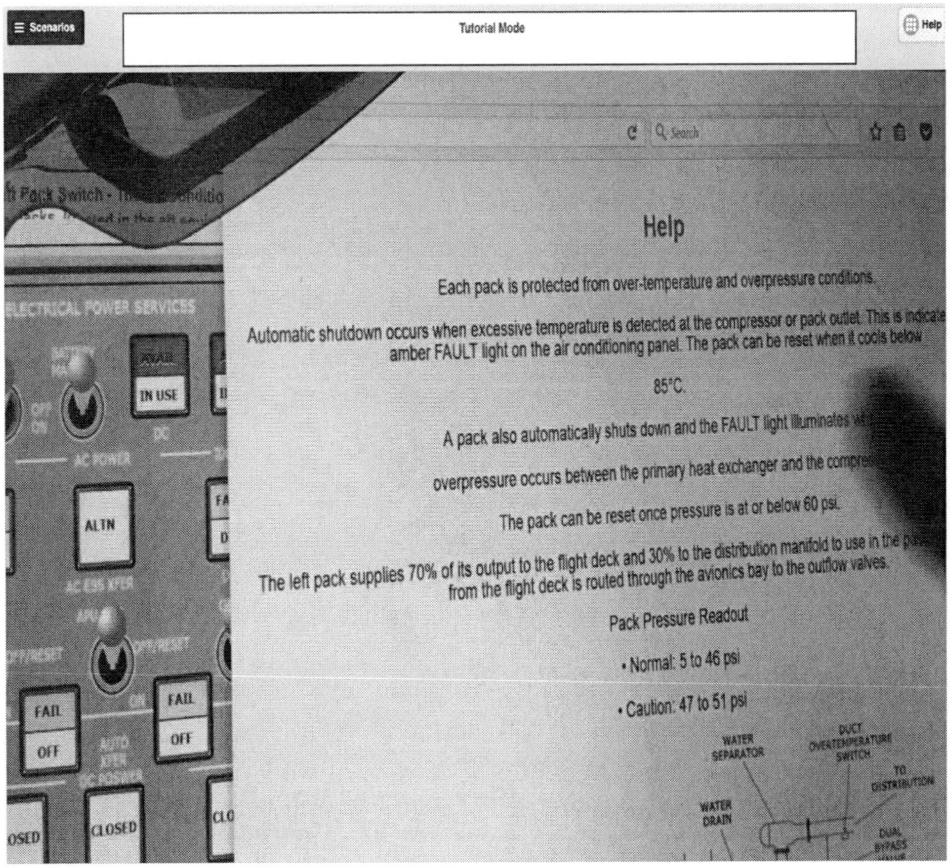

No More One size fits

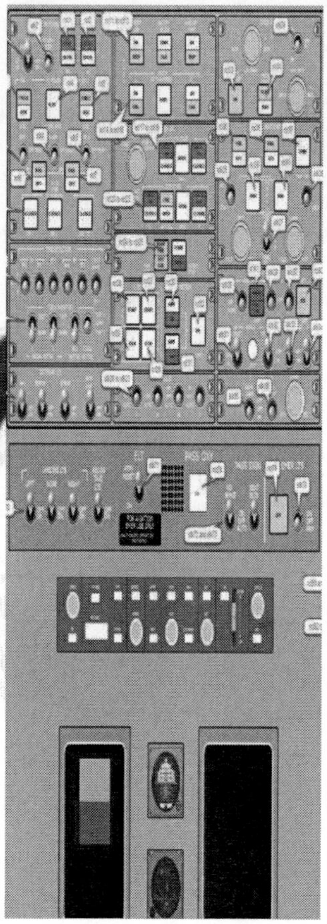

User defined training platforms

- Update as your manuals change
- Create your own training scenaric
- Integrate your company culture
- Integrate your procedures
- Upload your checklists, images an
- Embed AR 3D models
- Create the cockpit to match your
- Change with your changing needs

ADD a AR viewer or QR codes

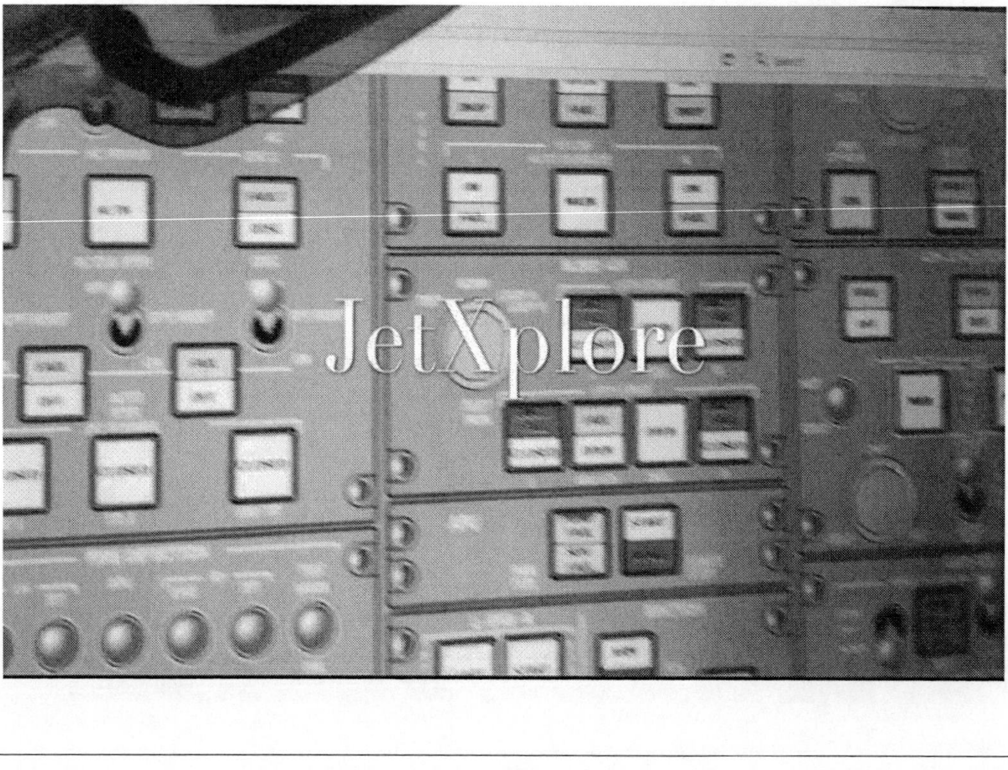

AUGMENT
Back to live view
Share
Hide shadows
Favo

Study and Explore anywhere

Overlay 3D Models on real equipment

Make your own QR Codes for training

ADD .QR to the shortened URL, paste in browser and you will have Code

Holo Lens

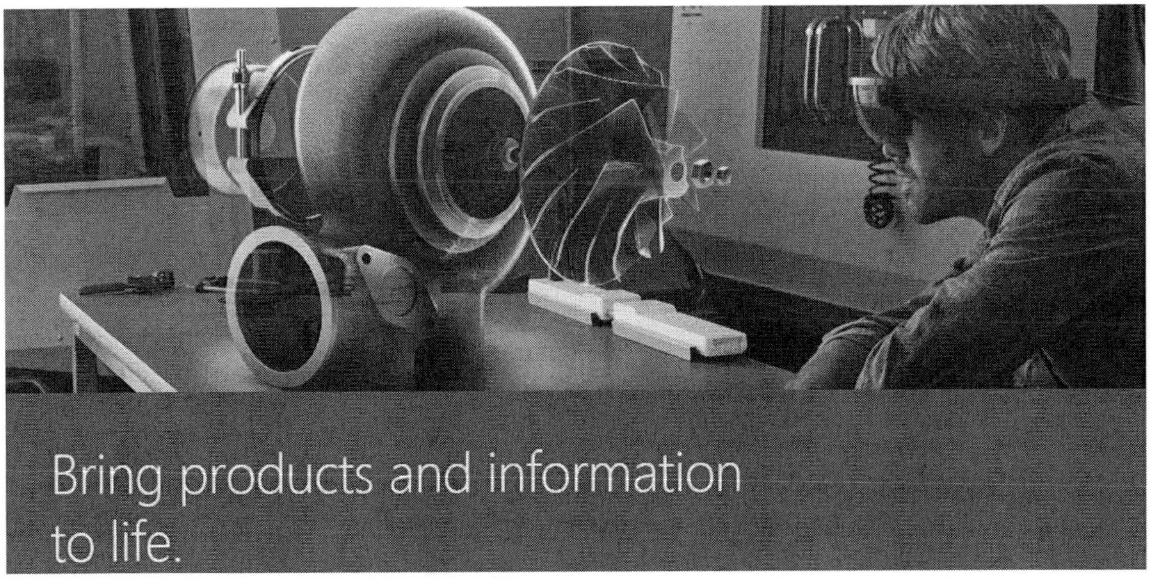

Bring products and information to life.

Thank You

Lori.Brown@wmich.edu

AVSOFT INTERNATIONAL
eLearning World Services
www.avsoft.com
A Perfect SCORM
by
Kim Halat

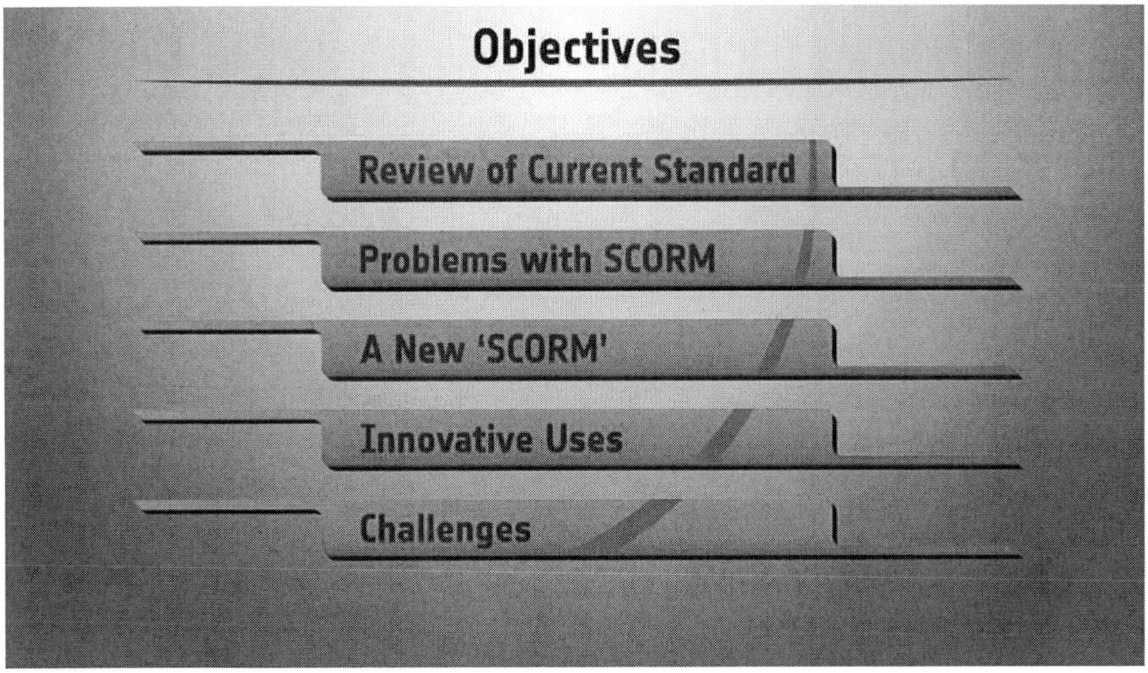
Objectives
Review of Current Standard
Problems with SCORM
A New 'SCORM'
Innovative Uses
Challenges

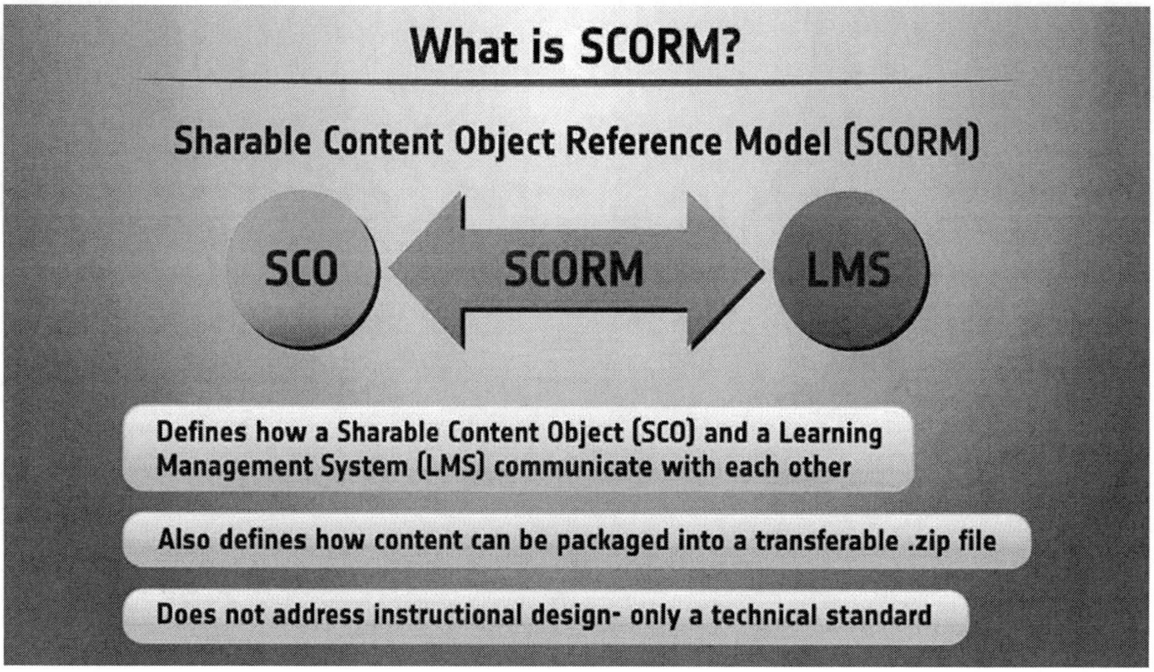
What is SCORM?
Sharable Content Object Reference Model (SCORM)
SCO
SCORM
LMS
Defines how a Sharable Content Object (SCO) and a Learning Management System (LMS) communicate with each other
Also defines how content can be packaged into a transferable .zip file
Does not address instructional design- only a technical standard

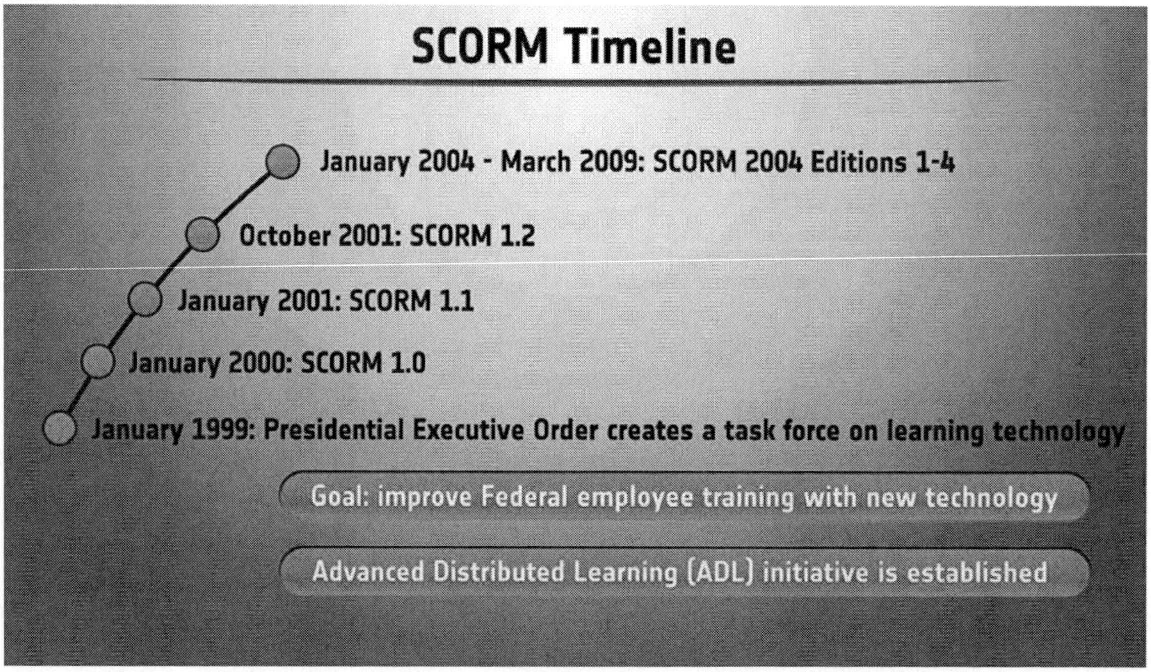
SCORM Timeline
January 2004 - March 2009: SCORM 2004 Editions 1-4
October 2001: SCORM 1.2
January 2001: SCORM 1.1
January 2000: SCORM 1.0
January 1999: Presidential Executive Order creates a task force on learning technology
Goal: improve Federal employee training with new technology
Advanced Distributed Learning (ADL) initiative is established

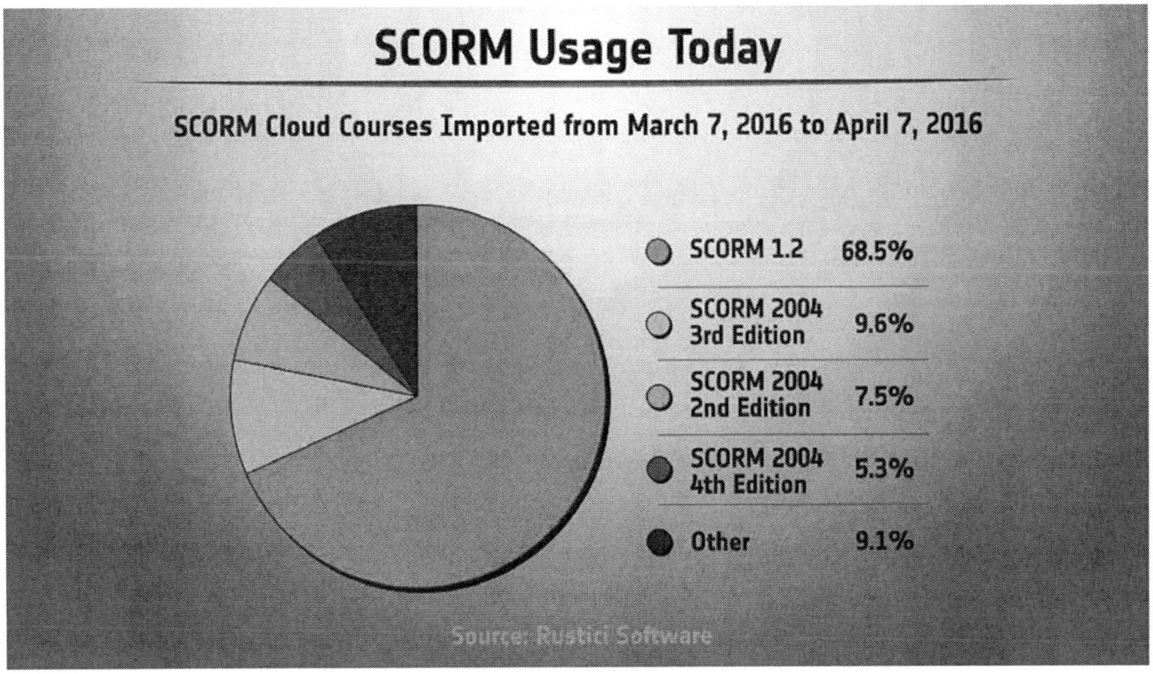
SCORM Usage Today
SCORM Cloud Courses Imported from March 7, 2016 to April 7, 2016
SCORM 1.2 68.5%
SCORM 2004 3rd Edition 9.6%
SCORM 2004 2nd Edition 7.5%
SCORM 2004 4th Edition 5.3%
Other 9.1%
Source: Rustici Software

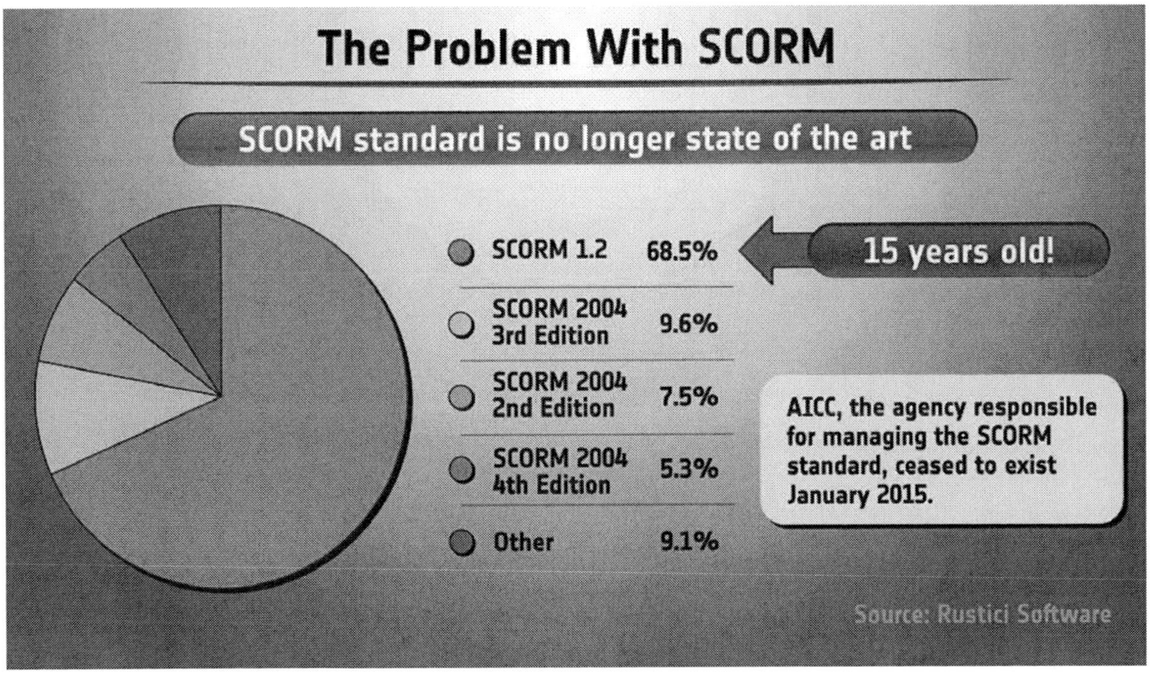
The Problem With SCORM
SCORM standard is no longer state of the art
SCORM 1.2 68.5%
15 years old!
SCORM 2004 3rd Edition 9.6%
SCORM 2004 2nd Edition 7.5%
SCORM 2004 4th Edition 5.3%
Other 9.1%
AICC, the agency responsible for managing the SCORM standard, ceased to exist January 2015.
Source: Rustici Software

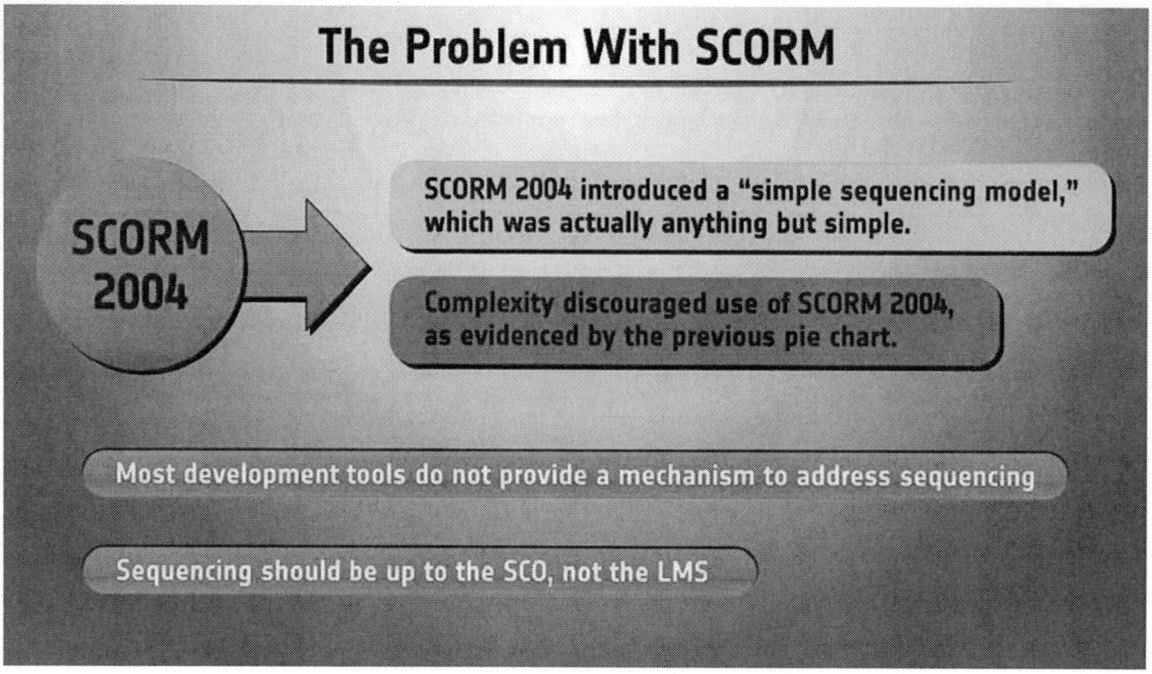
The Problem With SCORM
SCORM 2004
SCORM 2004 introduced a "simple sequencing model," which was actually anything but simple.
Complexity discouraged use of SCORM 2004, as evidenced by the previous pie chart.
Most development tools do not provide a mechanism to address sequencing
Sequencing should be up to the SCO, not the LMS

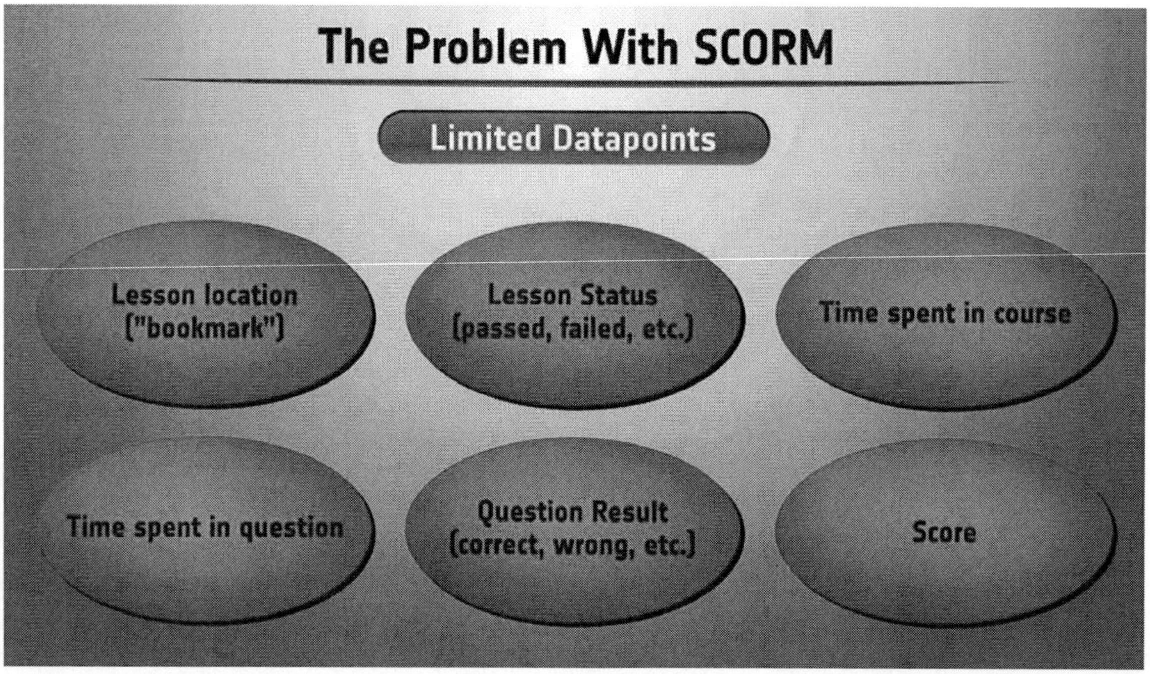
The Problem With SCORM
Limited Datapoints
Lesson location ("bookmark")
Lesson Status (passed, failed, etc.)
Time spent in course
Time spent in question
Question Result (correct, wrong, etc.)
Score

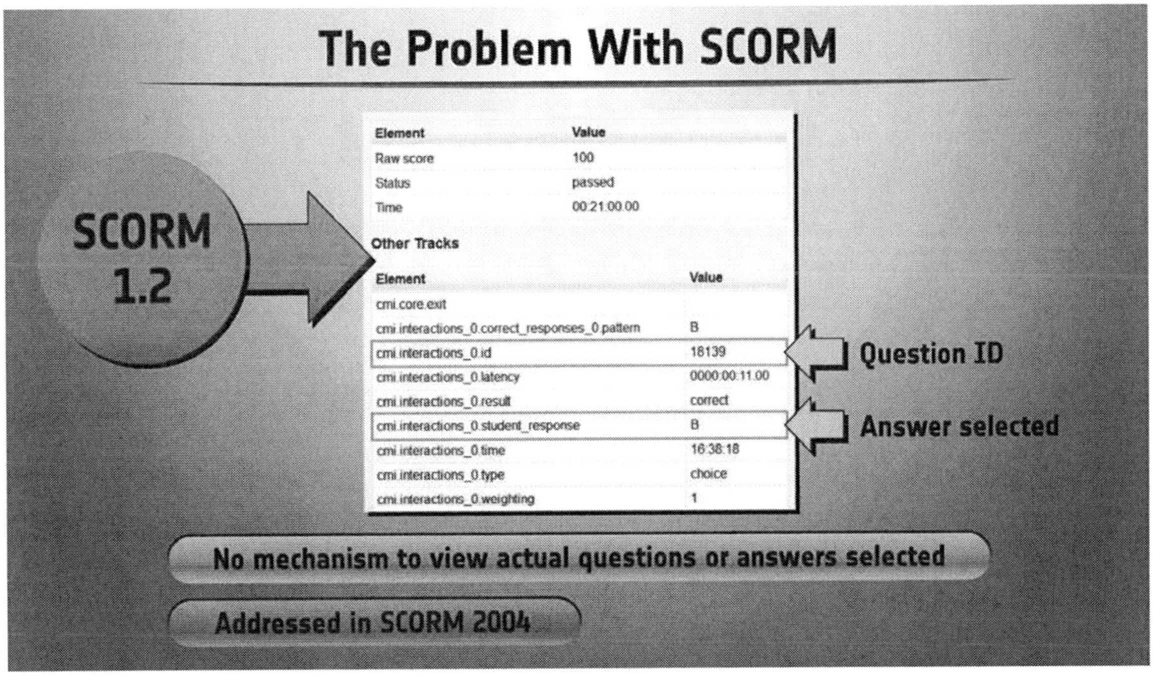
The Problem With SCORM
SCORM 1.2
Element
Value
Raw score
100
Status
passed
Time
00:21:00.00
Other Tracks
Element
Value
cmi.core.exit
cmi.interactions_0.correct_responses_0.pattern
B
cmi.interactions_0.id
18139
cmi.interactions_0.latency
0000:00:11.00
cmi.interactions_0.result
correct
cmi.interactions_0.student_response
B
cmi.interactions_0.time
16:38:18
cmi.interactions_0.type
choice
cmi.interactions_0.weighting
1
Question ID
Answer selected
No mechanism to view actual questions or answers selected
Addressed in SCORM 2004

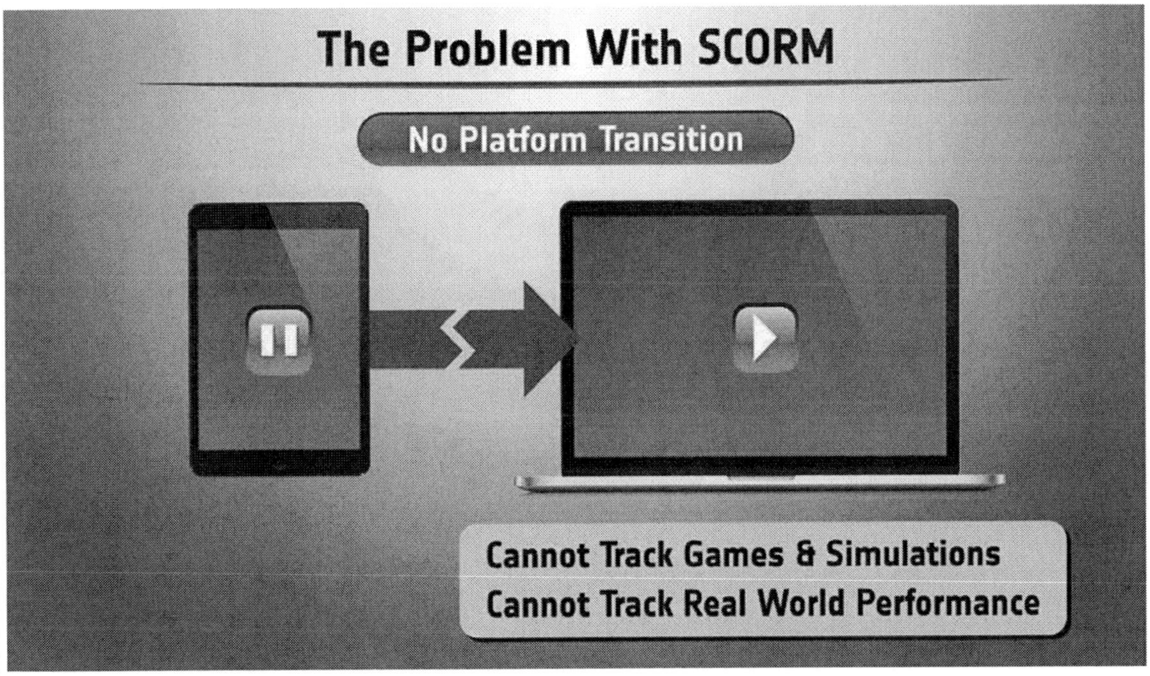
The Problem With SCORM
No Platform Transition
Cannot Track Games & Simulations
Cannot Track Real World Performance

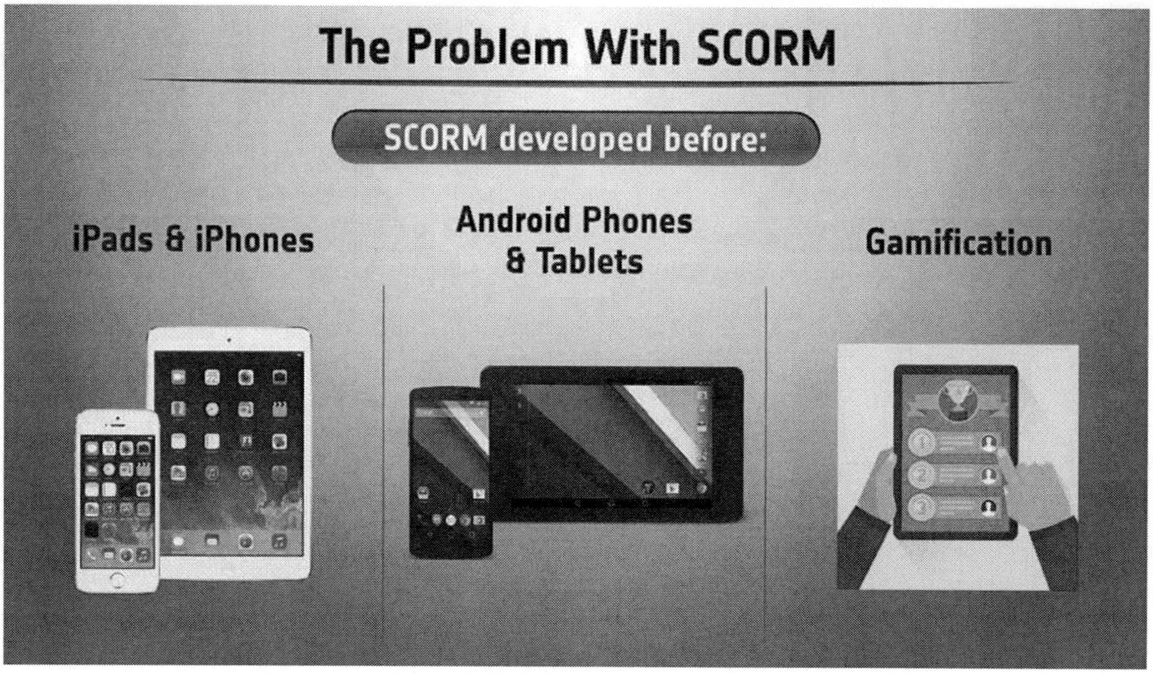
The Problem With SCORM
SCORM developed before:
iPads & iPhones
Android Phones
& Tablets
Gamification

The Problem With SCORM
SCORM standard does not work well in today's tech environment
SCORM
Browser
LMS
Internet
JavaScript

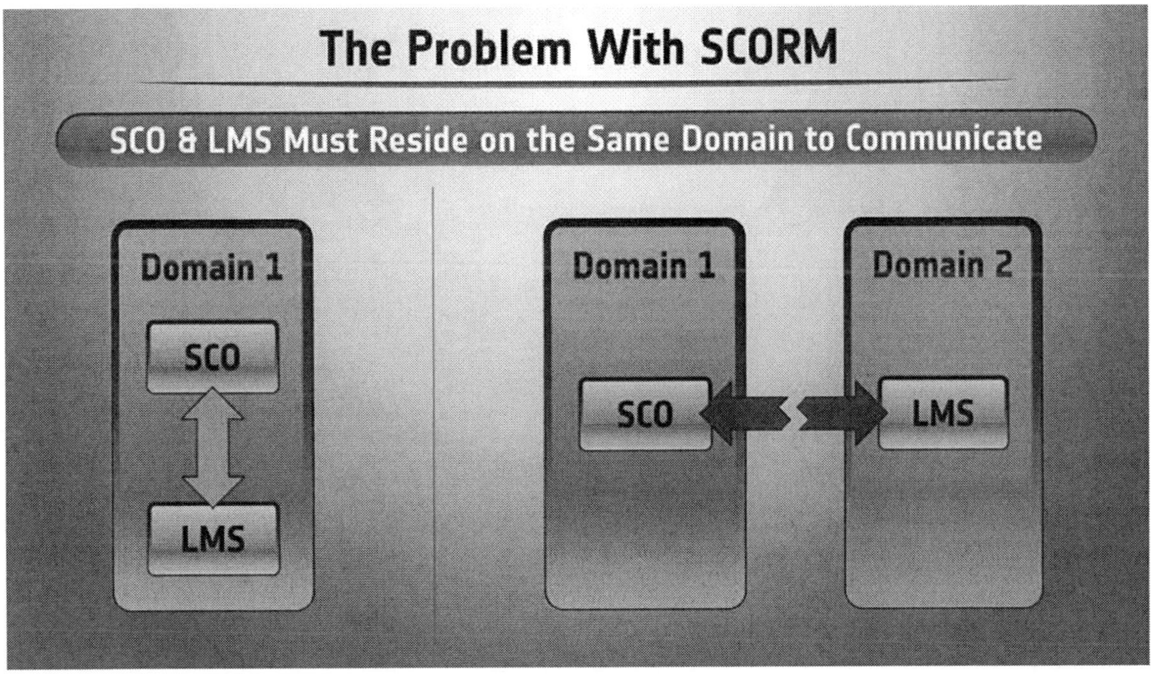

The Problem With SCORM
SCO & LMS Must Reside on the Same Domain to Communicate
Domain 1
SCO
LMS
Domain 1
SCO
Domain 2
LMS

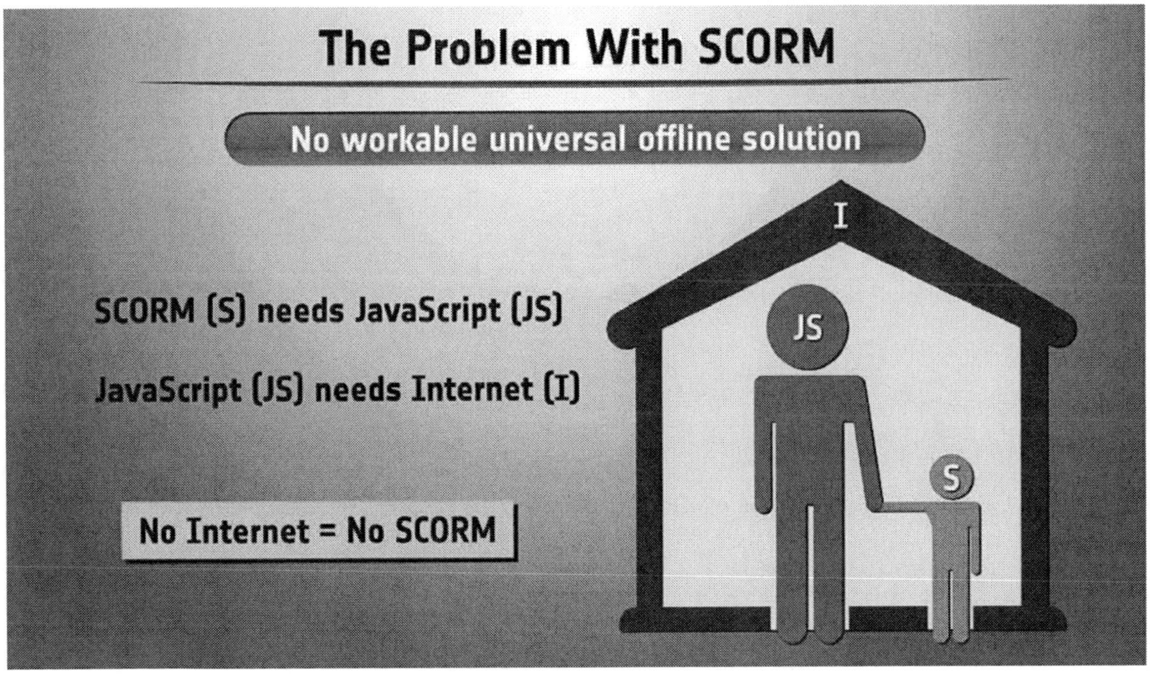

The Problem With SCORM
No workable universal offline solution
SCORM (S) needs JavaScript (JS)
JavaScript (JS) needs Internet (I)
No Internet = No SCORM
I
JS
S

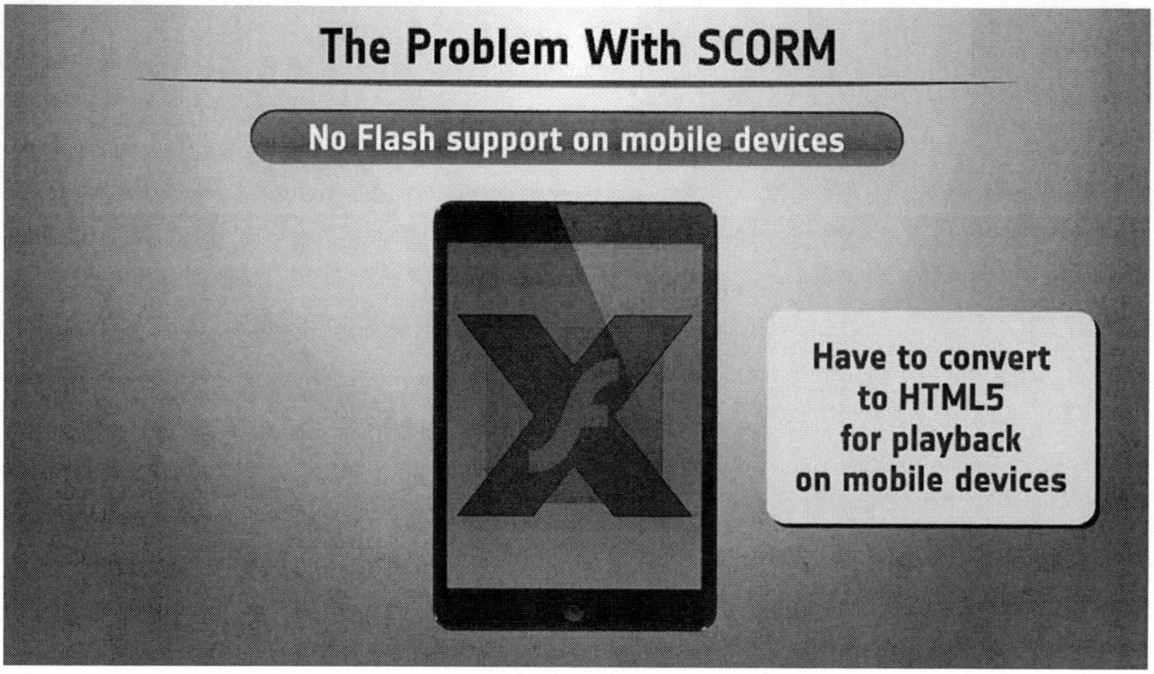
The Problem With SCORM
No Flash support on mobile devices
Have to convert
to HTML5
for playback
on mobile devices

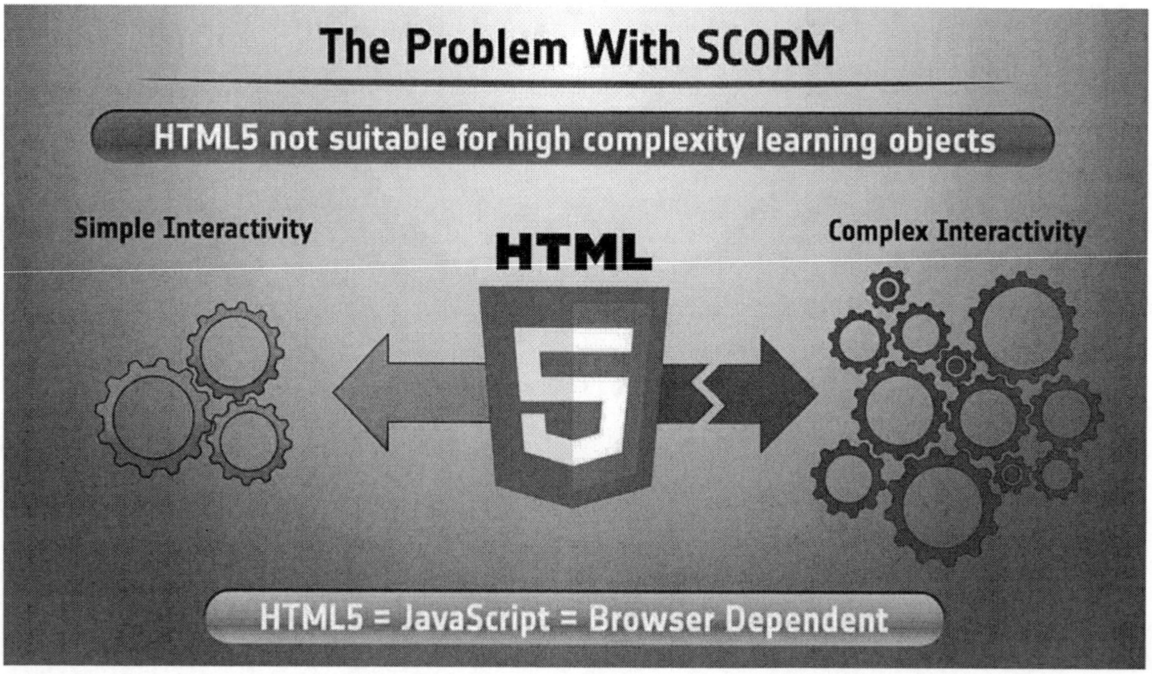
The Problem With SCORM
HTML5 not suitable for high complexity learning objects
Simple Interactivity
HTML
5
Complex Interactivity
HTML5 = JavaScript = Browser Dependent

The "Tin Can" Idea
Platform independent two-way communication protocol

The New SCORM
xAPI
April 2013: Version 1.0.0 is released, name changed to "Experience API" or "xAPI"
September 2011: first draft of new standard, called the "Tin Can API," is released
October 2010: ADL awards basic and applied research to Rustici Software to develop a new standard
API = Application Programming Interface = code that allows two software programs to communicate with each other

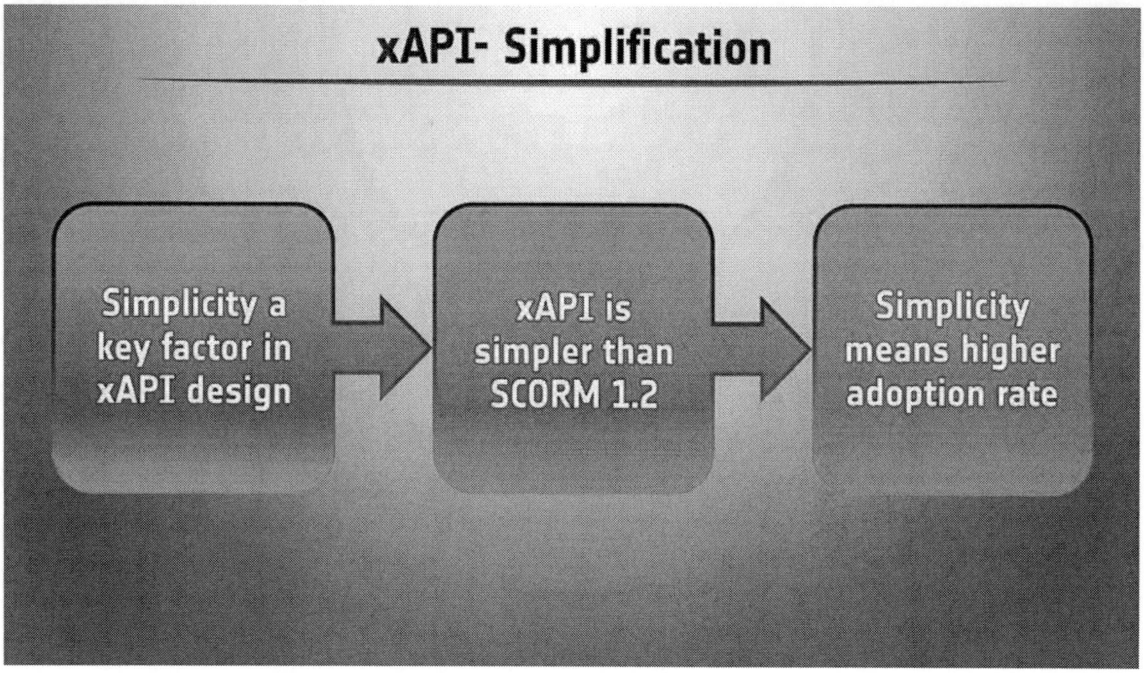
xAPI- Simplification
Simplicity a key factor in xAPI design
xAPI is simpler than SCORM 1.2
Simplicity means higher adoption rate

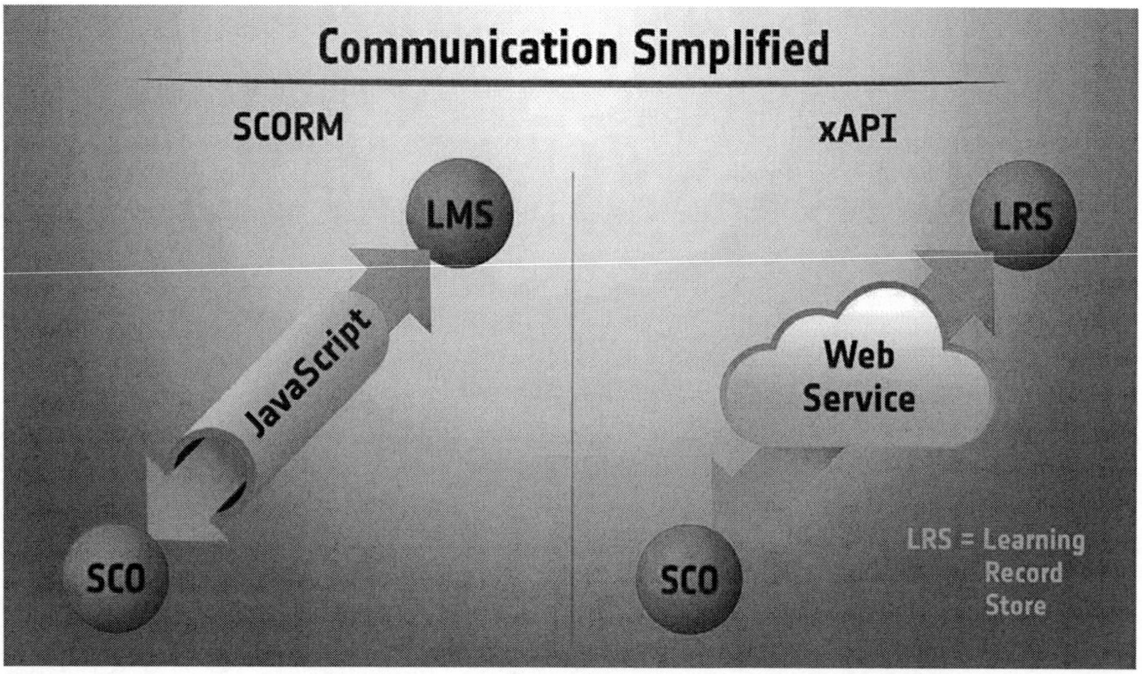
Communication Simplified
SCORM
xAPI
LMS
JavaScript
SCO
LRS
Web Service
SCO
LRS = Learning Record Store

What is a Web Service?

A piece of software that performs a specific task

Transmits data over the internet and allows programmatic access to that data

"Ask, and you shall receive"

General search

Local weather

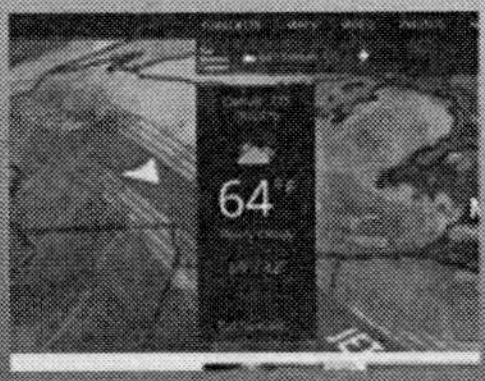

Travel search

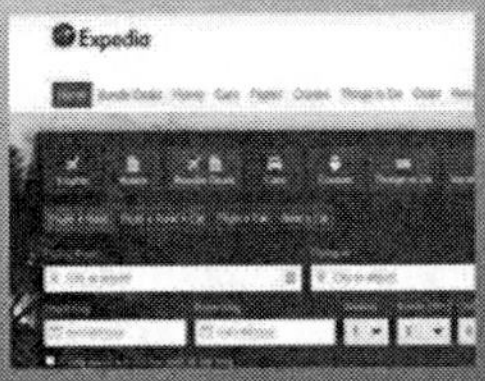

xAPI

xAPI- Technologically Agnostic

- Use in any training device ✓
- No restriction on programming language ✓
- Does not require JavaScript ✓
- Use online or offline ✓
- "Can your offline solution work with our system?" ✓
- Offers more freedom to content providers ✓

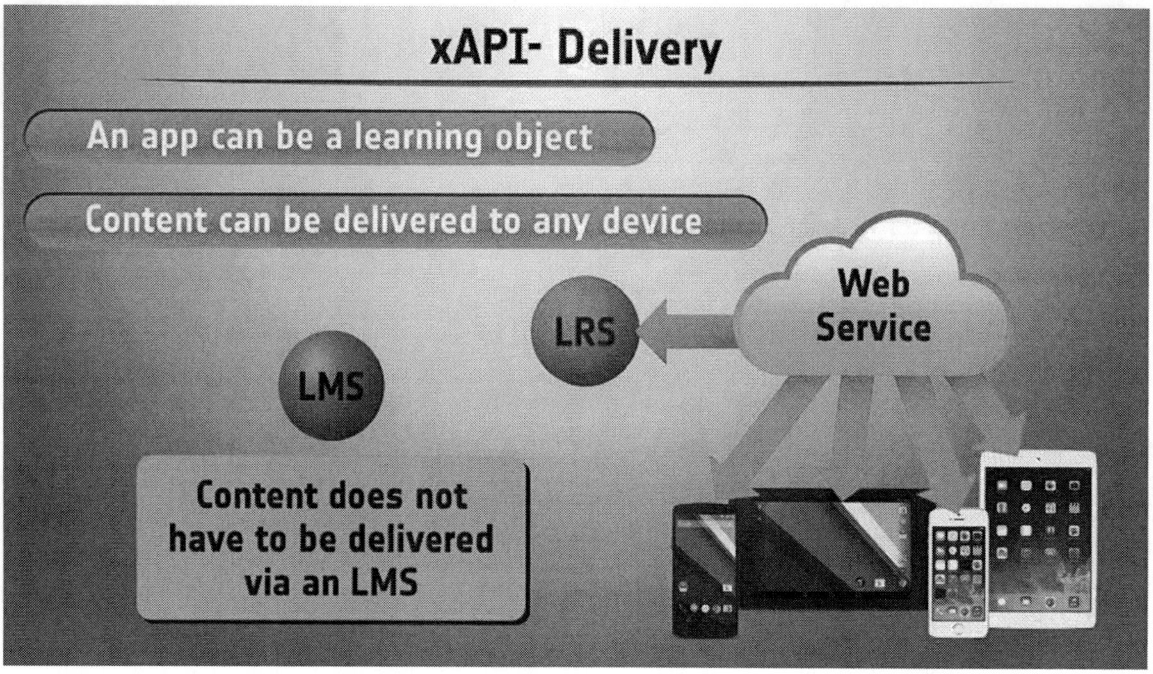

SCORM versus xAPI

	SCORM	xAPI
Track completion	✓	✓
Track time	✓	✓
Track pass/fail	✓	✓
Report a single score	✓	✓
Report multiple scores	✗	✓
Detailed test results	✗	✓
Solid security	✗	✓
No LMS required	✗	✓
No internet browser required	✗	✓
Keep complete content control	✗	✓
No cross-domain limitation	✗	✓
Use mobile apps for learning	✗	✓

	SCORM	xAPI
Platform transition	✗	✓
Track serious games	✗	✓
Track simulations	✗	✓
Track informal learning	✗	✓
Track real-world performance	✗	✓
Track offline learning	✗	✓
Track interactive learning	✗	✓
Track adaptive learning	✗	✓
Track blended learning	✗	✓
Track long-term learning	✗	✓
Track team-based learning	✗	✓

Source: http://experienceapi.com/scorm-vs-the-tin-can-api/

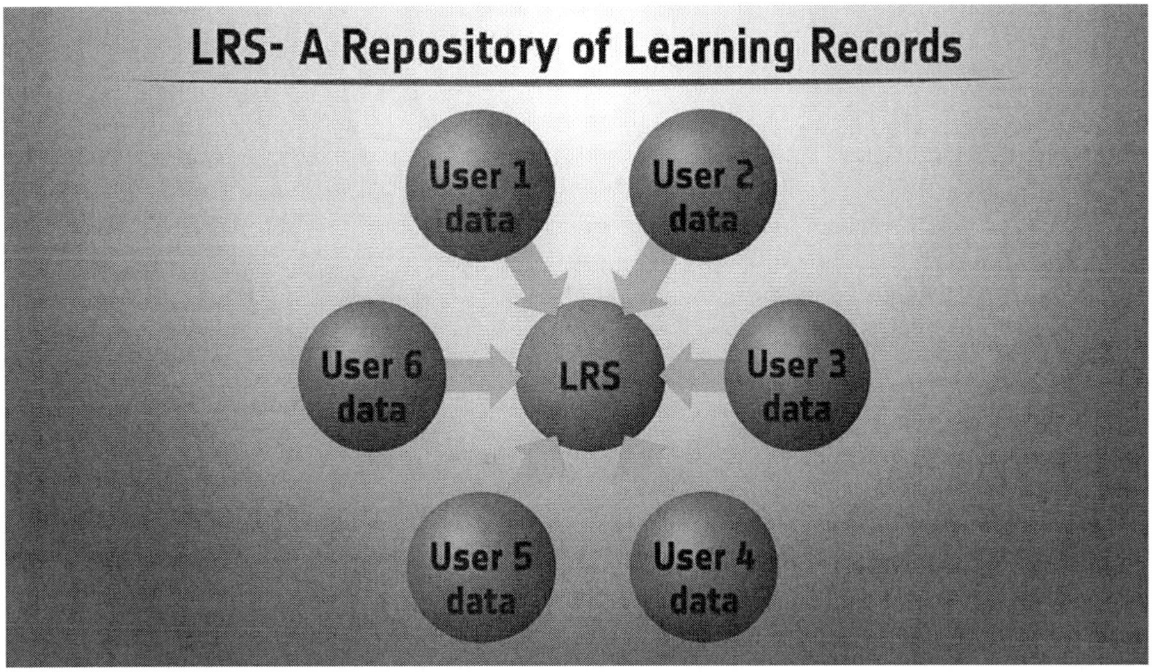
LRS- A Repository of Learning Records
User 1 data
User 2 data
User 6 data
LRS
User 3 data
User 5 data
User 4 data

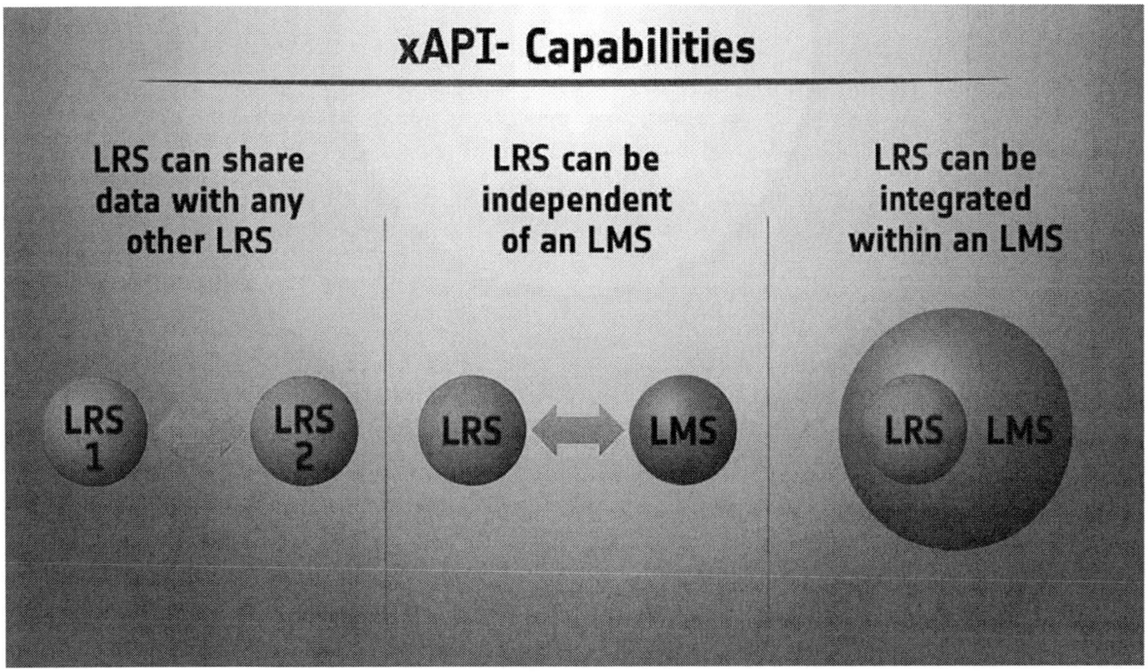
xAPI- Capabilities
LRS can share data with any other LRS
LRS can be independent of an LMS
LRS can be integrated within an LMS
LRS 1
LRS 2
LRS
LMS
LRS
LMS

xAPI- Big Data

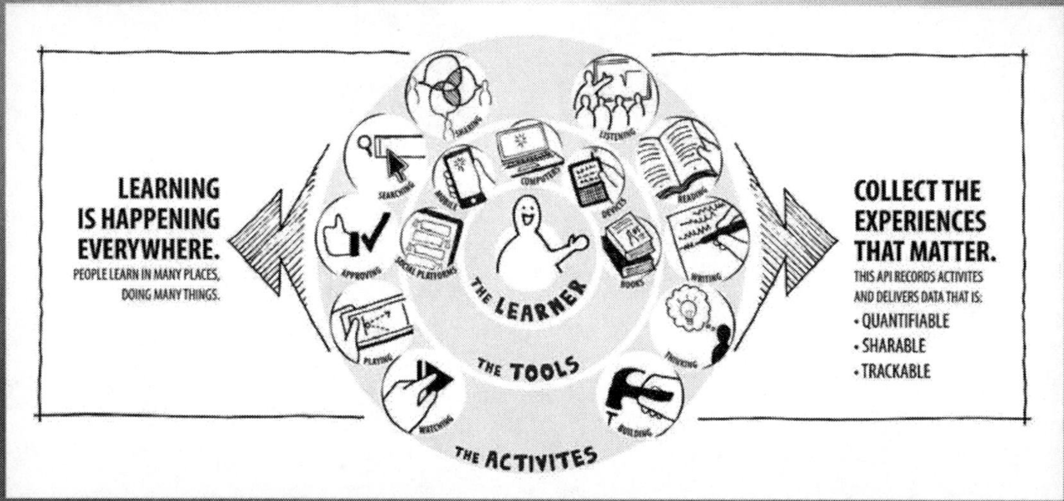

Source: http://tincanapi.com/overview/

xAPI- Statements

xAPI records user actions with "Actor, Verb, Object" statements

"Actor, Verb, Object" → **"Kim completed preflight"**

This allows almost any user activity to be recorded.

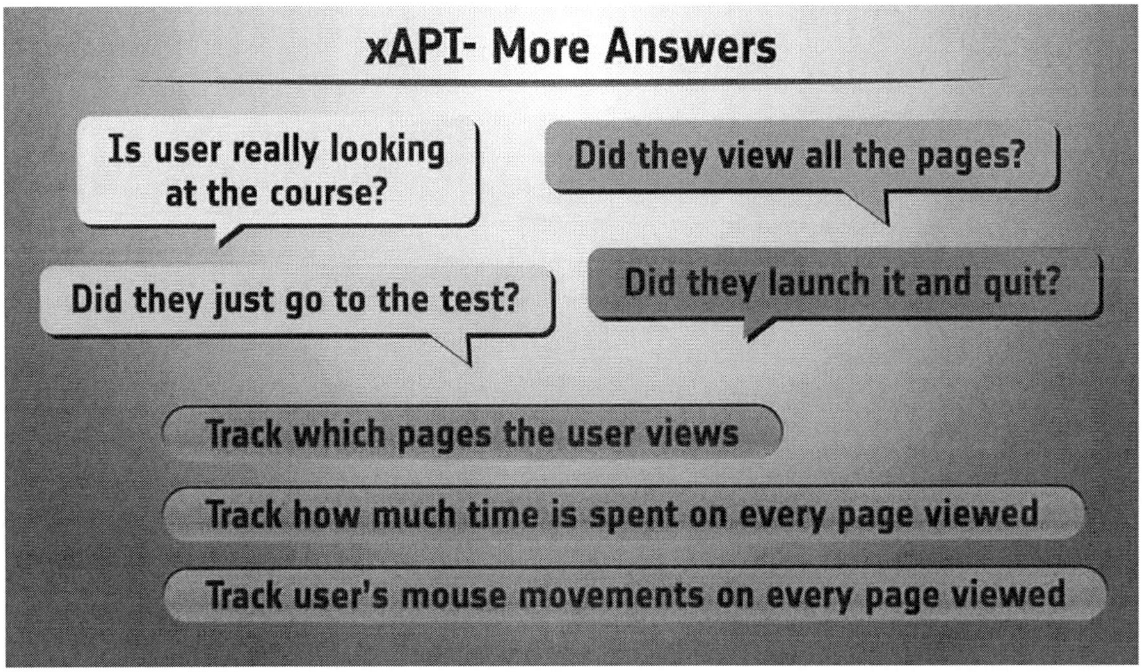
xAPI- More Answers
Is user really looking at the course?
Did they view all the pages?
Did they just go to the test?
Did they launch it and quit?
Track which pages the user views
Track how much time is spent on every page viewed
Track user's mouse movements on every page viewed

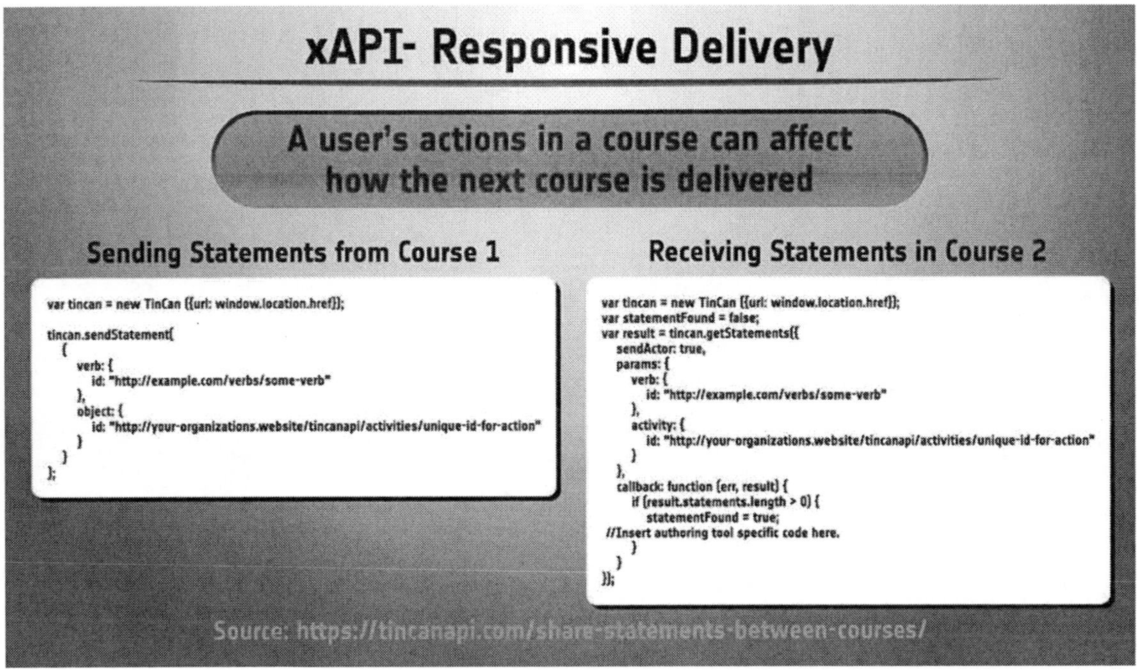
xAPI- Responsive Delivery
A user's actions in a course can affect how the next course is delivered
Sending Statements from Course 1
var tincan = new TinCan ({url: window.location.href});
tincan.sendStatement(
{
verb: {
id: "http://example.com/verbs/some-verb"
},
object: {
id: "http://your-organizations.website/tincanapi/activities/unique-id-for-action"
}
}
);
Receiving Statements in Course 2
var tincan = new TinCan ({url: window.location.href});
var statementFound = false;
var result = tincan.getStatements({
sendActor: true,
params: {
verb: {
id: "http://example.com/verbs/some-verb"
},
activity: {
id: "http://your-organizations.website/tincanapi/activities/unique-id-for-action"
}
},
callback: function (err, result) {
if (result.statements.length > 0) {
statementFound = true;
//Insert authoring tool specific code here.
}
}
});
Source: https://tincanapi.com/share-statements-between-courses/

xAPI- REAPER Project

1. Soldier shoots target
2. Target reports data to LRS
3. LRS triggers process to provide soldier with data visualization & training tips

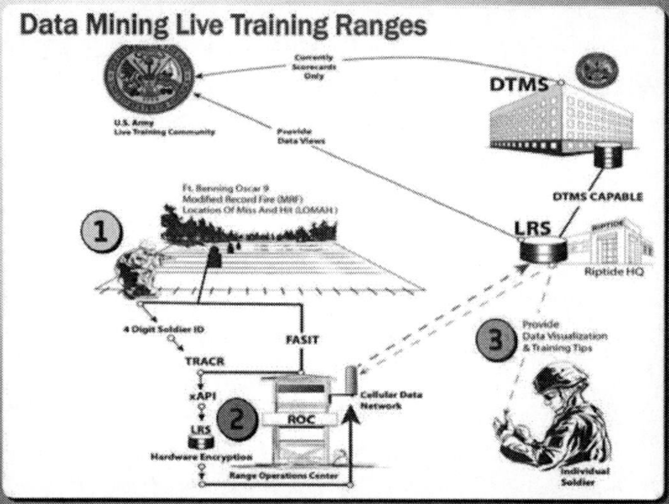

Source: http://www.slideshare.net/jhaag75/experience-api-xapi-and-the-future-of-scorm

xAPI- CPR Training

RISC chip measures user's actions to CPR manikin

e.g.,
- Location of compression
- Depth of compressions
- Rate of compressions
- Rate of ventilations
- Compression to ventilation ratio

RISC chip sends data to LRS

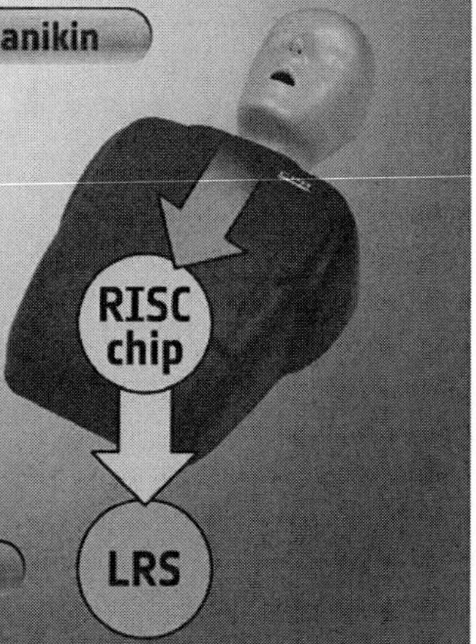

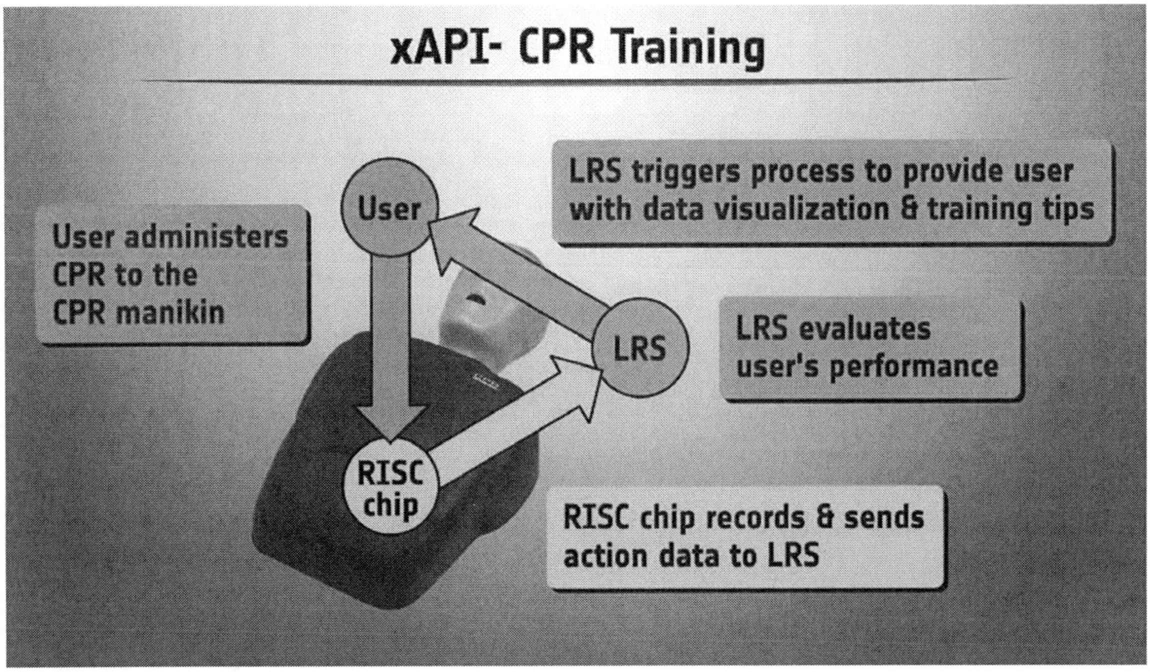
xAPI- CPR Training
User administers CPR to the CPR manikin
User
LRS triggers process to provide user with data visualization & training tips
LRS
LRS evaluates user's performance
RISC chip
RISC chip records & sends action data to LRS

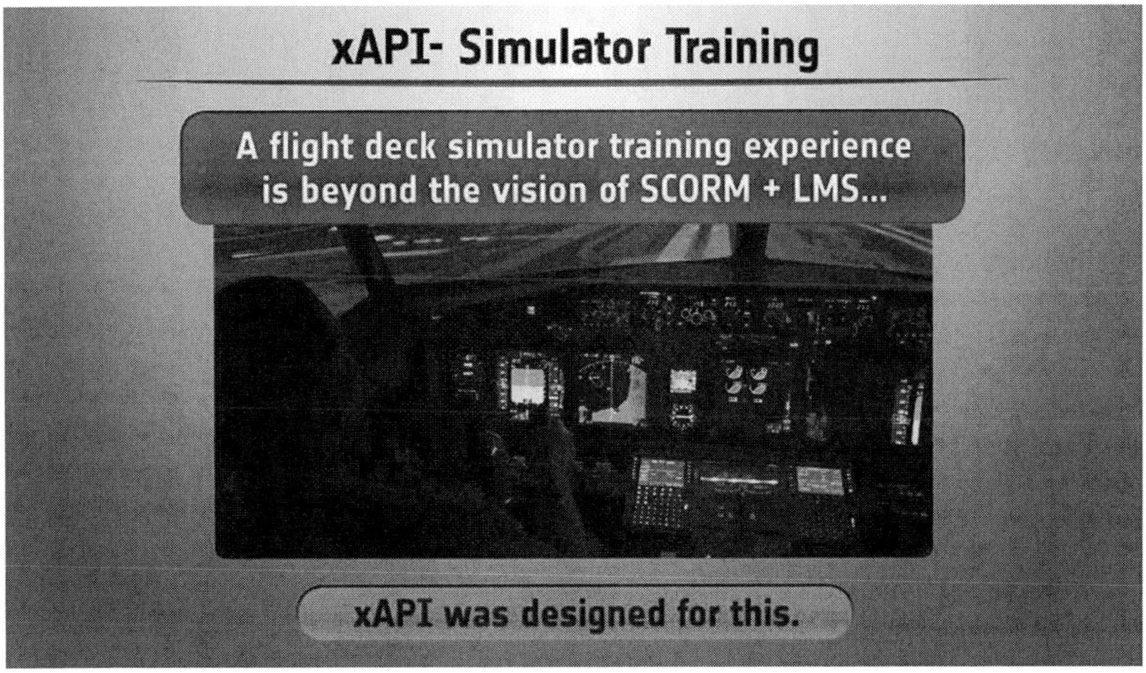
xAPI- Simulator Training
A flight deck simulator training experience is beyond the vision of SCORM + LMS...
xAPI was designed for this.

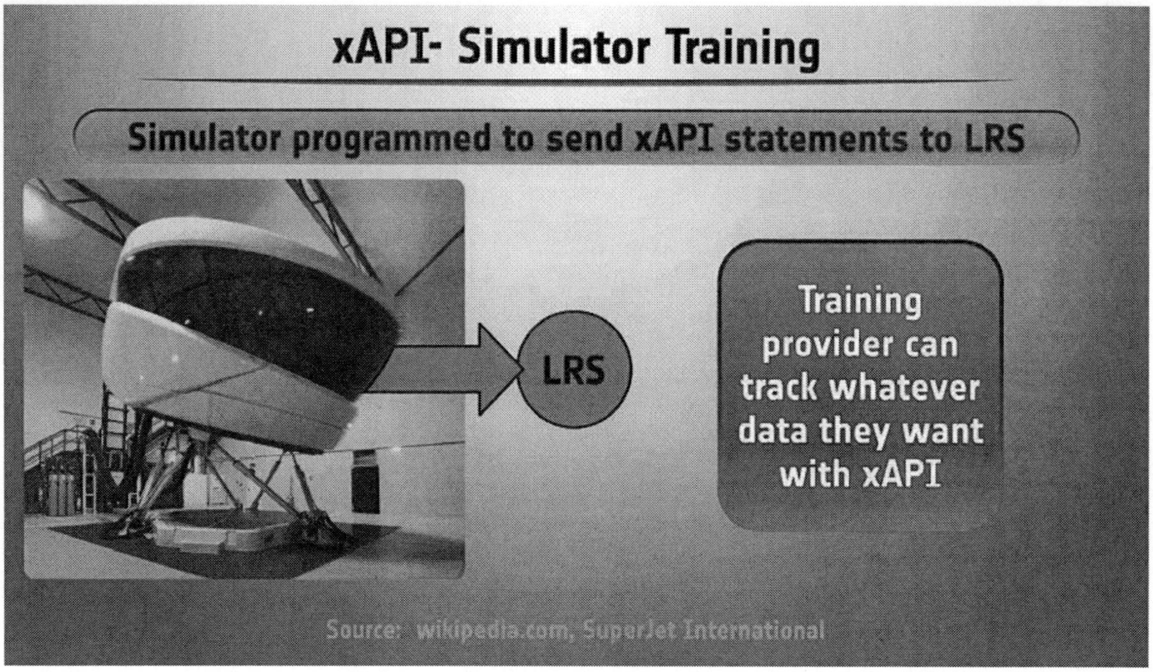
xAPI- Simulator Training
Simulator programmed to send xAPI statements to LRS
LRS
Training provider can track whatever data they want with xAPI
Source: wikipedia.com, SuperJet International

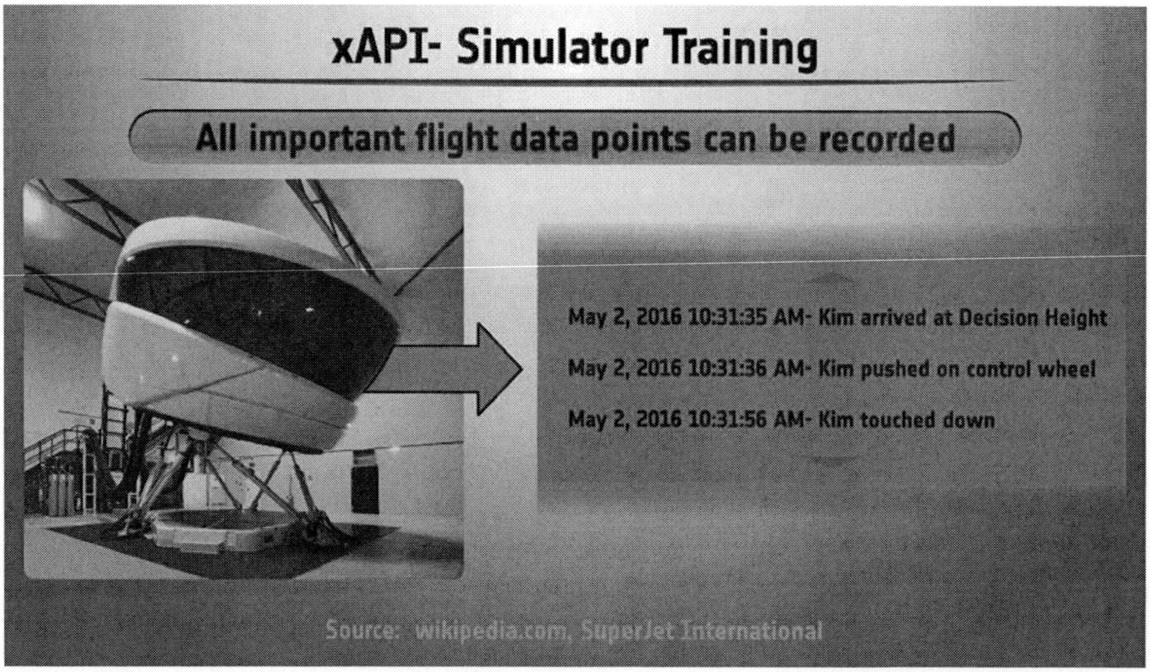
xAPI- Simulator Training
All important flight data points can be recorded
May 2, 2016 10:31:35 AM- Kim arrived at Decision Height
May 2, 2016 10:31:36 AM- Kim pushed on control wheel
May 2, 2016 10:31:56 AM- Kim touched down
Source: wikipedia.com, SuperJet International

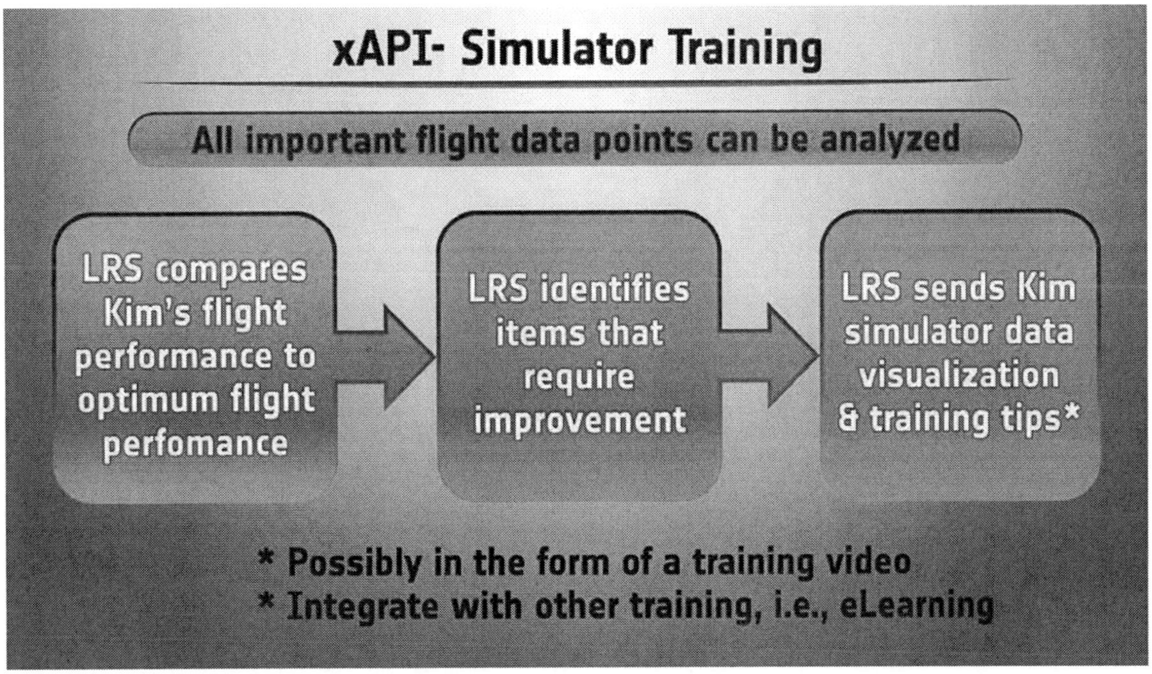
xAPI- Simulator Training
All important flight data points can be analyzed
LRS compares Kim's flight performance to optimum flight perfomance
LRS identifies items that require improvement
LRS sends Kim simulator data visualization & training tips*
* Possibly in the form of a training video
* Integrate with other training, i.e., eLearning

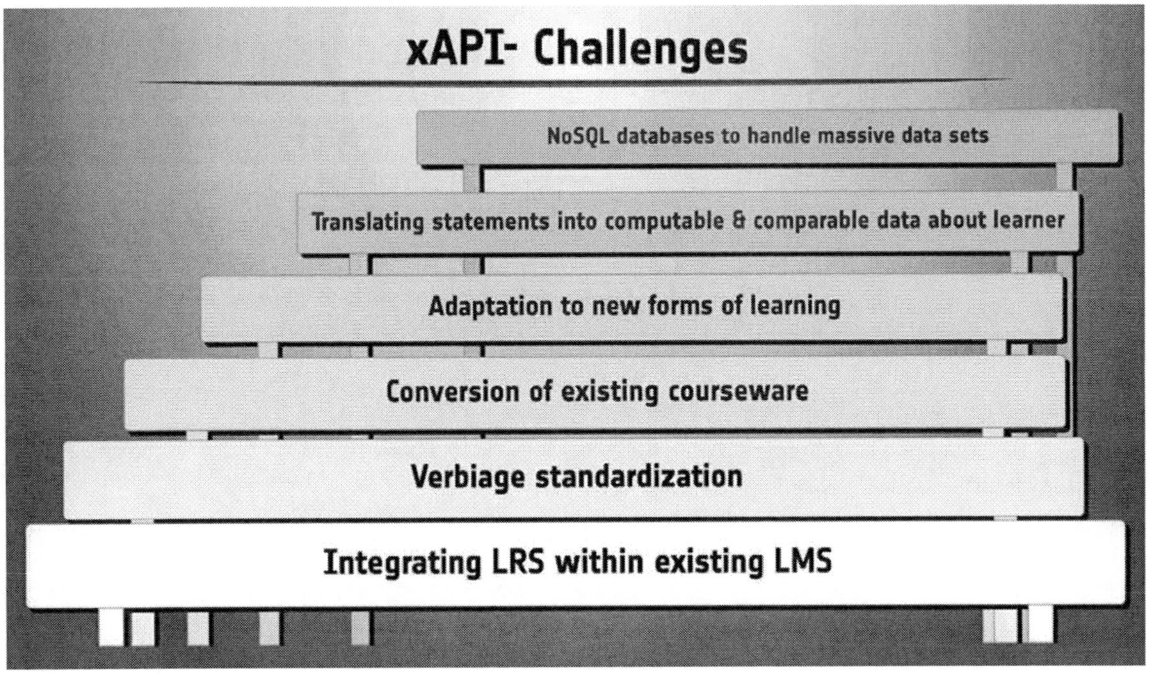
xAPI- Challenges
NoSQL databases to handle massive data sets
Translating statements into computable & comparable data about learner
Adaptation to new forms of learning
Conversion of existing courseware
Verbiage standardization
Integrating LRS within existing LMS

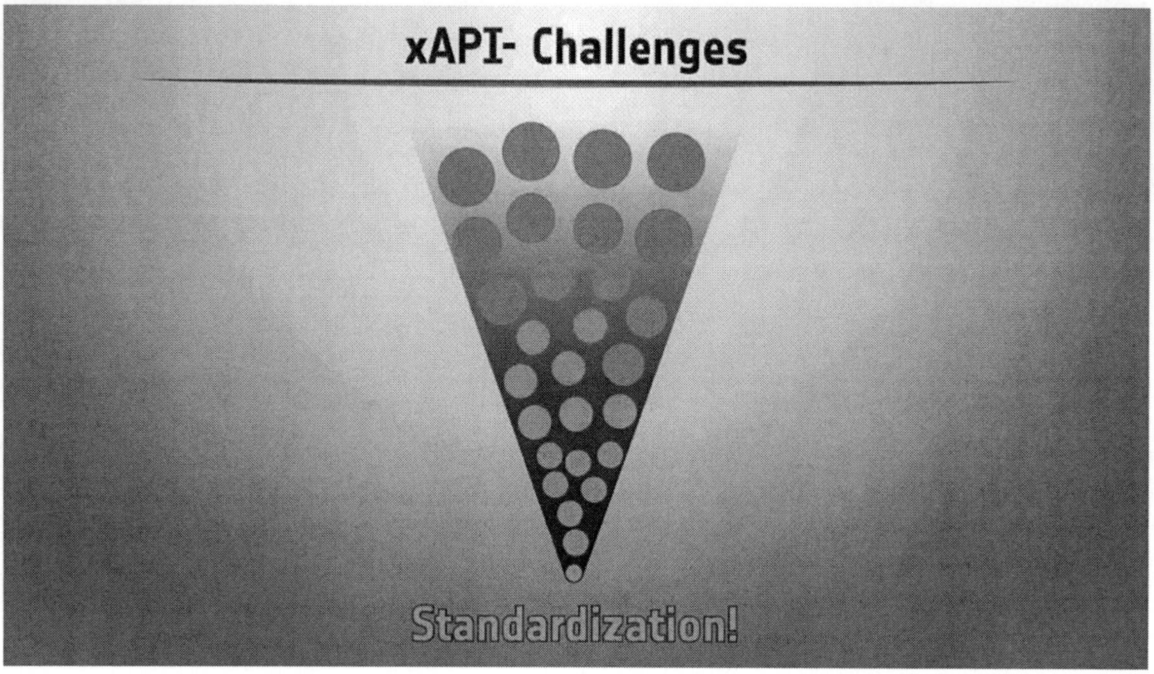
xAPI- Challenges
Standardization!

AVSOFT
INTERNATIONAL
eLearning World Services
www.avsoft.com
Thank You
Kim Halat
k_halat@avsoft.net
Sunrise over Placida Harbor, Florida

AUTHOR INDEX

Halldale Media Inc.
735 Primera Blvd.
Suite 220
Lake Mary, Florida 32746
USA

ISBN 978-1-5108-2541-3

CURRAN ASSOCIATES INC.
proceedings.com

XII Jornadas Iberoamericanas de Ingenieria de Software e Ingenieria del Conocimiento 2017 (JIISIC'17)

Held jointly with the Ecuadorian Conference on Software Engineering (CEIS'2017) and the Conference on Software Engineering Applied to Control and Automation Systems (ISASCA'2017)

Latacunga, Ecuador
17 – 18 August 2017